Thinking in Systems

Other Books by Donella H. Meadows:

Harvesting One Hundredfold: Key Concepts and Case Studies in Environmental Education (1989).
The Global Citizen (1991).

with Dennis Meadows:
Toward Global Equilibrium (1973).

with Dennis Meadows and Jørgen Randers:
Beyond the Limits (1992).
Limits to Growth: The 30-Year Update (2004).

with Dennis Meadows, Jørgen Randers, and William W. Behrens III:
The Limits to Growth (1972).

with Dennis Meadows, et al.:
The Dynamics of Growth in a Finite World (1974).

with J. Richardson and G. Bruckmann:
Groping in the Dark: The First Decade of Global Modeling (1982).

with J. Robinson:
The Electronic Oracle: Computer Models and Social Decisions (1985).

Thinking in Systems

— A Primer —

Donella H. Meadows

Edited by Diana Wright,
Sustainability Institute

CHELSEA GREEN PUBLISHING
WHITE RIVER JUNCTION, VERMONT

Project Manager: Emily Foote
Developmental Editor: Joni Praded
Copy Editor: Cannon Labrie
Proofreader: Ellen Brownstein
Indexer: Beth Nauman-Montana
Designer: Peter Holm, Sterling Hill Productions

Printed in the United States of America
First printing, December, 2008
10 9 8 7 6 5 4   10 11 12

Our Commitment to Green Publishing
Chelsea Green sees publishing as a tool for cultural change and ecological stewardship. We strive to align our book manufacturing practices with our editorial mission and to reduce the impact of our business enterprise in the environment. We print our books and catalogs on chlorine-free recycled paper, using soy-based inks whenever possible. This book may cost slightly more because we use recycled paper, and we hope you'll agree that it's worth it. Chelsea Green is a member of the Green Press Initiative (www.greenpressinitiative.org), a nonprofit coalition of publishers, manufacturers, and authors working to protect the world's endangered forests and conserve natural resources. *Thinking in Systems* was printed on 55-lb. Natures Book Natural, a 30-percent postconsumer-waste, FSC-certified, recycled paper supplied by Thomson-Shore.

Library of Congress Cataloging-in-Publication Data
    Meadows, Donella H.
      Thinking in systems : a primer / Donella H. Meadows ; edited by Diana Wright.
        p. cm.
      Includes bibliographical references.
      ISBN 978-1-60358-055-7
      1. System analysis--Simulation methods 2. Decision making--Simulation methods 3. Critical thinking--Simulation methods 4. Sustainable development--Simulation methods. 5. Social sciences--Simulation methods. 6. Economic development--Environmental aspects--Simulation methods. 7. Population--Economic aspects--Simulation methods. 8. Pollution--Economic aspects--Simulation methods. 9. Environmental education--Simulation methods. I. Wright, Diana. II. Title.
      QA402.M425 2008
      003--dc22

                                        2008035211

Chelsea Green Publishing Company
Post Office Box 428
White River Junction, VT 05001
(802) 295-6300
www.chelseagreen.com

Part of this work has been adapted from an article originally published under the title "Whole Earth Models and Systems" in *Coevolution Quarterly* (Summer 1982). An early version of Chapter 6 appeared as "Places to Intervene in a System" in *Whole Earth Review* (Winter 1997) and later as an expanded paper published by the Sustainability Institute. Chapter 7, "Living in a World of Systems," was originally published as "Dancing with Systems" in *Whole Earth Review* (Winter 2001).

FOR DANA
(1941–2001)

and for all those who would learn from her

CONTENTS

A NOTE FROM THE AUTHOR

This book has been distilled out of the wisdom of thirty years of systems modeling and teaching carried out by dozens of creative people, most of them originally based at or influenced by the MIT System Dynamics group. Foremost among them is Jay Forrester, the founder of the group. My particular teachers (and students who have become my teachers) have been, in addition to Jay: Ed Roberts, Jack Pugh, Dennis Meadows, Hartmut Bossel, Barry Richmond, Peter Senge, John Sterman, and Peter Allen, but I have drawn here from the language, ideas, examples, quotes, books, and lore of a large intellectual community. I express my admiration and gratitude to all its members.

I also have drawn from thinkers in a variety of disciplines, who, as far as I know, never used a computer to simulate a system, but who are natural systems thinkers. They include Gregory Bateson, Kenneth Boulding, Herman Daly, Albert Einstein, Garrett Hardin, Václav Havel, Lewis Mumford, Gunnar Myrdal, E.F. Schumacher, a number of modern corporate executives, and many anonymous sources of ancient wisdom, from Native Americans to the Sufis of the Middle East. Strange bedfellows, but systems thinking transcends disciplines and cultures and, when it is done right, it overarches history as well.

Having spoken of transcendence, I need to acknowledge factionalism as well. Systems analysts use overarching concepts, but they have entirely human personalities, which means that they have formed many fractious schools of systems thought. I have used the language and symbols of system dynamics here, the school in which I was taught. And I present only the core of systems theory here, not the leading edge. I don't deal with the most abstract theories and am interested in analysis only when I can see how it helps solve real problems. When the abstract end of systems theory does that, which I believe it will some day, another book will have to be written.

Therefore, you should be warned that this book, like all books, is biased and incomplete. There is much, much more to systems thinking than is presented here, for you to discover if you are interested. One of my purposes is to make you interested. Another of my purposes, the main one, is to give you a basic ability to understand and to deal with complex systems, even if your formal systems training begins and ends with this book.

—DONELLA MEADOWS, 1993

A NOTE FROM THE EDITOR

In 1993, Donella (Dana) Meadows completed a draft of the book you now hold. The manuscript was not published at the time, but circulated informally for years. Dana died quite unexpectedly in 2001—before she completed this book. In the years since her death, it became clear that her writings have continued to be useful to a wide range of readers. Dana was a scientist and writer, and one of the best communicators in the world of systems modeling.

In 1972, Dana was lead author of *The Limits to Growth*—a best-selling and widely translated book. The cautions she and her fellow authors issued then are recognized today as the most accurate warnings of how unsustainable patterns could, if unchecked, wreak havoc across the globe. That book made headlines around the world for its observations that continual growth in population and consumption could severely damage the ecosystems and social systems that support life on earth, and that a drive for limitless economic growth could eventually disrupt many local, regional, and global systems. The findings in that book and its updates are, once again, making front-page news as we reach peak oil, face the realities of climate change, and watch a world of 6.6 billion people deal with the devastating consequences of physical growth.

In short, Dana helped usher in the notion that we have to make a major shift in the way we view the world and its systems in order to correct our course. Today, it is widely accepted that systems thinking is a critical tool in addressing the many environmental, political, social, and economic challenges we face around the world. Systems, big or small, can behave in similar ways, and understanding those ways is perhaps our best hope for making lasting change on many levels. Dana was writing this book to bring that concept to a wider audience, and that is why I and my colleagues at the Sustainability Institute decided it was time to publish her manuscript posthumously.

Will another book really help the world and help you, the reader? I think so. Perhaps you are working in a company (or own a company) and are struggling to see how your business or organization can be part of a shift toward a better world. Or maybe you're a policy maker who is seeing others "push back" against your good ideas and good intentions. Perhaps you're a manager who has worked hard to fix some important problems in your company or community, only to see other challenges erupt in their wake. As one who advocates for changes in how a society (or a family) functions, what it values and protects, you may see years of progress easily undone in a few swift reactions. As a citizen of an increasingly global society, perhaps you are just plain frustrated with how hard it is to make a positive and lasting difference.

If so, I think that this book can help. Although one can find dozens of titles on "systems modeling" and "systems thinking," there remains a clear need for an approachable and inspiring book about systems and us—why we find them at times so baffling and how we can better learn to manage and redesign them.

At the time that Dana was writing *Thinking in Systems*, she had recently completed the twenty-year update to *Limits to Growth*, titled *Beyond the Limits*. She was a Pew Scholar in Conservation and the Environment, was serving on the Committee on Research and Exploration for the National Geographic Society, and she was teaching about systems, environment, and ethics at Dartmouth College. In all aspects of her work, she was immersed in the events of the day. She understood those events to be the outward behavior of often complex systems.

Although Dana's original manuscript has been edited and restructured, many of the examples you will find in this book are from her first draft in 1993. They may seem a bit dated to you, but in editing her work I chose to keep them because their teachings are as relevant now as they were then. The early 1990s were the time of the dissolution of the Soviet Union and great shifts in other socialist countries. The North American Free Trade Agreement was newly signed. Iraq's army invaded Kuwait and then retreated, burning oil fields on the way out. Nelson Mandela was freed from prison, and South Africa's apartheid laws were repealed. Labor leader Lech Walesa was elected president of Poland, and poet Václav Havel was elected president of Czechoslovakia. The International Panel on Climate Change issued its first assessment report, concluding that "emissions from human activities are substantially increasing the atmospheric concentrations of greenhouse gases and that this will enhance the greenhouse effect and result in an additional warming of the Earth's surface." The UN held a conference in Rio de Janeiro on environment and development.

While traveling to meetings and conferences during this time, Dana read the *International Herald Tribune* and during a single week found many examples of systems in need of better management or complete redesign. She found them in the newspaper because they are all around us every day. Once you start to see the events of the day as parts of trends, and those trends as symptoms of underlying system structure, you will be able to consider new ways to manage and new ways to live in a world of complex systems. In publishing Dana's manuscript, I hope to increase the ability of readers to understand and talk about the systems around them and to act for positive change.

I hope this small approachable introduction to systems and how we think about them will be a useful tool in a world that rapidly needs to shift behaviors arising from very complex systems. This is a simple book for and about a complex world. It is a book for those who want to shape a better future.

—DIANA WRIGHT, 2008

If a factory is torn down but the rationality which produced it is left standing, then that rationality will simply produce another factory. If a revolution destroys a government, but the systematic patterns of thought that produced that government are left intact, then those patterns will repeat themselves. . . . There's so much talk about the system. And so little understanding.

—ROBERT PIRSIG, *Zen and the Art of Motorcycle Maintenance*

Introduction: The System Lens

Managers are not confronted with problems that are independent of each other, but with dynamic situations that consist of complex systems of changing problems that interact with each other. I call such situations messes. . . . Managers do not solve problems, they manage messes.

—RUSSELL ACKOFF,[1] operations theorist

Early on in teaching about systems, I often bring out a Slinky. In case you grew up without one, a Slinky is a toy—a long, loose spring that can be made to bounce up and down, or pour back and forth from hand to hand, or walk itself downstairs.

I perch the Slinky on one upturned palm. With the fingers of the other hand, I grasp it from the top, partway down its coils. Then I pull the bottom hand away. The lower end of the Slinky drops, bounces back up again, yo-yos up and down, suspended from my fingers above.

"What made the Slinky bounce up and down like that?" I ask students.

"Your hand. You took away your hand," they say.

So I pick up the box the Slinky came in and hold it the same way, poised on a flattened palm, held from above by the fingers of the other hand. With as much dramatic flourish as I can muster, I pull the lower hand away.

Nothing happens. The box just hangs there, of course.

"Now once again. What made the Slinky bounce up and down?"

The answer clearly lies within the Slinky itself. The hands that manipulate it suppress or release some behavior that is latent within the structure of the spring.

That is a central insight of systems theory.

Once we see the relationship between structure and behavior, we can begin to understand how systems work, what makes them produce poor results, and how to shift them into better behavior patterns. As our world continues to change rapidly and become more complex, systems thinking will help us manage, adapt, and see the wide range of choices we have before us. It is a way of thinking that gives us the freedom to identify root causes of problems and see new opportunities.

So, what is a system? A system is a set of things—people, cells, molecules, or whatever—interconnected in such a way that they produce their own pattern of behavior over time. The system may be buffeted, constricted, triggered, or driven by outside forces. But the system's response to these forces is characteristic of itself, and that response is seldom simple in the real world.

When it comes to Slinkies, this idea is easy enough to understand. When it comes to individuals, companies, cities, or economies, it can be heretical. The system, to a large extent, causes its own behavior! An outside event may unleash that behavior, but the same outside event applied to a different system is likely to produce a different result.

Think for a moment about the implications of that idea:

- Political leaders don't cause recessions or economic booms. Ups and downs are inherent in the structure of the market economy.
- Competitors rarely cause a company to lose market share. They may be there to scoop up the advantage, but the losing company creates its losses at least in part through its own business policies.
- The oil-exporting nations are not solely responsible for oil-price rises. Their actions alone could not trigger global price rises and economic chaos if the oil consumption, pricing, and investment policies of the oil-importing nations had not built economies that are vulnerable to supply interruptions.
- The flu virus does not attack you; you set up the conditions for it to flourish within you.
- Drug addiction is not the failing of an individual and no one person, no matter how tough, no matter how loving, can cure a drug addict—not even the addict. It is only through understanding addiction as part of a larger set of influences and societal issues that one can begin to address it.

Something about statements like these is deeply unsettling. Something else is purest common sense. I submit that those two somethings—a resistance to and a recognition of systems principles—come from two kinds of human experience, both of which are familiar to everyone.

On the one hand, we have been taught to analyze, to use our rational ability, to trace direct paths from cause to effect, to look at things in small and understandable pieces, to solve problems by acting on or controlling the world around us. That training, the source of much personal and societal power, leads us to see presidents and competitors, OPEC and the flu and drugs as the causes of our problems.

On the other hand, long before we were educated in rational analysis, we all dealt with complex systems. We are complex systems—our own bodies are magnificent examples of integrated, interconnected, self-maintaining complexity. Every person we encounter, every organization, every animal, garden, tree, and forest is a complex system. We have built up intuitively, without analysis, often without words, a practical understanding of how these systems work, and how to work with them.

Modern systems theory, bound up with computers and equations, hides the fact that it traffics in truths known at some level by everyone. It is often possible, therefore, to make a direct translation from systems jargon to traditional wisdom.

> Because of feedback delays within complex systems, by the time a problem becomes apparent it may be unnecessarily difficult to solve.
> — *A stitch in time saves nine.*

> According to the competitive exclusion principle, if a reinforcing feedback loop rewards the winner of a competition with the means to win further competitions, the result will be the elimination of all but a few competitors.
> — *For he that hath, to him shall be given; and he that hath not, from him shall be taken even that which he hath (Mark 4:25)*
>   or
> —*The rich get richer and the poor get poorer.*

A diverse system with multiple pathways and redundancies is

more stable and less vulnerable to external shock than a uniform system with little diversity.

— *Don't put all your eggs in one basket.*

Ever since the Industrial Revolution, Western society has benefited from science, logic, and reductionism over intuition and holism. Psychologically and politically we would much rather assume that the cause of a problem is "out there," rather than "in here." It's almost irresistible to blame something or someone else, to shift responsibility away from ourselves, and to look for the control knob, the product, the pill, the technical fix that will make a problem go away.

Serious problems have been solved by focusing on external agents—preventing smallpox, increasing food production, moving large weights and many people rapidly over long distances. Because they are embedded in larger systems, however, some of our "solutions" have created further problems. And some problems, those most rooted in the internal structure of complex systems, the real messes, have refused to go away.

Hunger, poverty, environmental degradation, economic instability, unemployment, chronic disease, drug addiction, and war, for example, persist in spite of the analytical ability and technical brilliance that have been directed toward eradicating them. No one deliberately creates those problems, no one wants them to persist, but they persist nonetheless. That is because they are intrinsically systems problems—undesirable behaviors characteristic of the system structures that produce them. They will yield only as we reclaim our intuition, stop casting blame, see the system as the source of its own problems, and find the courage and wisdom to *restructure* it.

Obvious. Yet subversive. An old way of seeing. Yet somehow new. Comforting, in that the solutions are in our hands. Disturbing, because we must *do things*, or at least *see things* and *think about things*, in a different way.

This book is about that different way of seeing and thinking. It is intended for people who may be wary of the word "systems" and the field of systems analysis, even though they may have been doing systems thinking all their lives. I have kept the discussion nontechnical because I want to show what a long way you can go toward understanding systems without turning to mathematics or computers.

I have made liberal use of diagrams and time graphs in this book because there is a problem in discussing systems only with words. Words and sentences must, by necessity, come only one at a time in linear, logical order. Systems happen all at once. They are connected not just in one direction, but in many directions simultaneously. To discuss them properly, it is necessary somehow to use a language that shares some of the same properties as the phenomena under discussion.

Pictures work for this language better than words, because you can see all the parts of a picture at once. I will build up systems pictures gradually, starting with very simple ones. I think you'll find that you can understand this graphical language easily.

I start with the basics: the definition of a system and a dissection of its parts (in a reductionist, unholistic way). Then I put the parts back together to show how they interconnect to make the basic operating unit of a system: the feedback loop.

Next I will introduce you to a systems zoo—a collection of some common and interesting types of systems. You'll see how a few of these creatures behave and why and where they can be found. You'll recognize them; they're all around you and even within you.

With a few of the zoo "animals"—a set of specific examples—as a foundation, I'll step back and talk about how and why systems work so beautifully and the reasons why they so often surprise and confound us. I'll talk about why everyone or everything in a system can act dutifully and rationally, yet all these well-meaning actions too often add up to a perfectly terrible result. And why things so often happen much faster or slower than everyone thinks they will. And why you can be doing something that has always worked and suddenly discover, to your great disappointment, that your action no longer works. And why a system might suddenly, and without warning, jump into a kind of behavior you've never seen before.

That discussion will lead to us to look at the common problems that the systems-thinking community has stumbled upon over and over again through working in corporations and governments, economies and ecosystems, physiology and psychology. "There's another case of the tragedy of the commons," we find ourselves saying as we look at an allocation system for sharing water resource among communities or financial resources among schools. Or we identify "eroding goals" as we study the business rules and incentives that help or hinder the development of new technologies. Or we see "policy resistance" as we examine decision-making power and the nature of relationships in a family, a community, or a nation. Or we witness "addiction"—which can be caused by many more agents than caffeine, alcohol, nicotine, and narcotics.

Systems thinkers call these common structures that produce characteristic behaviors "archetypes." When I first planned this book, I called them "system traps." Then I added the words "and opportunities," because these archetypes, which are responsible for some of the most intransigent and potentially dangerous problems, also can be transformed, with a little systems understanding, to produce much more desirable behaviors.

From this understanding I move into what you and I can do about restructuring the systems we live within. We can learn how to look for leverage points for change.

I conclude with the largest lessons of all, the ones derived from the wisdom shared by most systems thinkers I know. For those who want to explore systems thinking further, the Appendix provides ways to dig deeper into the subject with a glossary, a bibliography of systems thinking resources, a summary list of systems principles, and equations for the models described in Part One.

When our small research group moved from MIT to Dartmouth College years ago, one of the Dartmouth engineering professors watched us in seminars for a while, and then dropped by our offices. "You people are different," he said. "You ask different kinds of questions. You see things I don't see. Somehow you come at the world in a different way. How? Why?"

That's what I hope to get across throughout this book, but especially in its conclusion. I don't think the systems way of seeing is better than the reductionist way of thinking. I think it's complementary, and therefore revealing. You can see some things through the lens of the human eye, other things through the lens of a microscope, others through the lens of a telescope, and still others through the lens of systems theory. Everything seen through each kind of lens is actually there. Each way of seeing allows our knowledge of the wondrous world in which we live to become a little more complete.

At a time when the world is more messy, more crowded, more interconnected, more interdependent, and more rapidly changing than ever before, the more ways of seeing, the better. The systems-thinking lens allows us to reclaim our intuition about whole systems and

• hone our abilities to understand parts,

- see interconnections,
- ask "what-if" questions about possible future behaviors, and
- be creative and courageous about system redesign.

Then we can use our insights to make a difference in ourselves and our world.

INTERLUDE • *The Blind Men and the Matter of the Elephant*

Beyond Ghor, there was a city. All its inhabitants were blind. A king with his entourage arrived nearby; he brought his army and camped in the desert. He had a mighty elephant, which he used to increase the people's awe.

The populace became anxious to see the elephant, and some sightless from among this blind community ran like fools to find it.

As they did not even know the form or shape of the elephant, they groped sightlessly, gathering information by touching some part of it.

Each thought that he knew something, because he could feel a part. . . .

The man whose hand had reached an ear . . . said: "It is a large, rough thing, wide and broad, like a rug."

And the one who had felt the trunk said: "I have the real facts about it. It is like a straight and hollow pipe, awful and destructive."

The one who had felt its feet and legs said: "It is mighty and firm, like a pillar."

Each had felt one part out of many. Each had perceived it wrongly. . . .[2]

This ancient Sufi story was told to teach a simple lesson but one that we often ignore: The behavior of a system cannot be known just by knowing the elements of which the system is made.

PART ONE

System Structure and Behavior

— ONE —

The Basics

I have yet to see any problem, however complicated, which, when
looked at in the right way, did not become still more complicated.

—Poul Anderson[1]

More Than the Sum of Its Parts

A system isn't just any old collection of things. A **system*** is an interconnected set of elements that is coherently organized in a way that achieves something. If you look at that definition closely for a minute, you can see that a system must consist of three kinds of things: *elements, interconnections*, and a *function* or *purpose*.

For example, the elements of your digestive system include teeth, enzymes, stomach, and intestines. They are interrelated through the physical flow of food, and through an elegant set of regulating chemical signals. The function of this system is to break down food into its basic nutrients and to transfer those nutrients into the bloodstream (another system), while discarding unusable wastes.

A football team is a system with elements such as players, coach, field, and ball. Its interconnections are the rules of the game, the coach's strategy, the players' communications, and the laws of physics that govern the motions of ball and players. The purpose of the team is to win games, or have fun, or get exercise, or make millions of dollars, or all of the above.

A school is a system. So is a city, and a factory, and a corporation, and a national economy. An animal is a system. A tree is a system, and a forest is a larger system that encompasses subsystems of trees and animals. The earth

---
* Definitions of words in bold face can be found in the Glossary.

is a system. So is the solar system; so is a galaxy. Systems can be embedded in systems, which are embedded in yet other systems.

Is there anything that is not a system? Yes—a conglomeration without any particular interconnections or function. Sand scattered on a road by happenstance is not, itself, a system. You can add sand or take away sand and you still have just sand on the road. Arbitrarily add or take away football players, or pieces of your digestive system, and you quickly no longer have the same system.

When a living creature dies, it loses its "system-ness." The multiple interrelations that held it together no longer function, and it dissipates, although its material remains part of a larger food-web system. Some people say that an old city neighborhood where people know each other and communicate regularly is a social system, and that a new apartment block full of strangers is not—not until new relationships arise and a system forms.

> **A system is more than the sum of its parts**. It may exhibit adaptive, dynamic, goal-seeking, self-preserving, and sometimes evolutionary behavior.

You can see from these examples that there is an integrity or wholeness about a system and an active set of mechanisms to maintain that integrity. Systems can change, adapt, respond to events, seek goals, mend injuries, and attend to their own survival in lifelike ways, although they may contain or consist of nonliving things. Systems can be self-organizing, and often are self-repairing over at least some range of disruptions. They are resilient, and many of them are evolutionary. Out of one system other completely new, never-before-imagined systems can arise.

Look Beyond the Players to the Rules of the Game

> You think that because you understand "one" that you must therefore understand "two" because one and one make two. But you forget that you must also understand "and."
>
> —Sufi teaching story

The elements of a system are often the easiest parts to notice, because many of them are visible, tangible things. The elements that make up a tree are roots, trunk, branches, and leaves. If you look more closely, you

---

THINK ABOUT THIS

How to know whether you are looking at a system or just a bunch of stuff:

A) Can you identify parts? . . . and

B) Do the parts affect each other? . . . and

C) Do the parts together produce an effect that is different from the effect of each part on its own? . . . and perhaps

D) Does the effect, the behavior over time, persist in a variety of circumstances?

---

see specialized cells: vessels carrying fluids up and down, chloroplasts, and so on. The system called a university is made up of buildings, students, professors, administrators, libraries, books, computers—and I could go on and say what all those things are made up of. Elements do not have to be physical things. Intangibles are also elements of a system. In a university, school pride and academic prowess are two intangibles that can be very important elements of the system. Once you start listing the elements of a system, there is almost no end to the process. You can divide elements into sub-elements and then sub-sub-elements. Pretty soon you lose sight of the system. As the saying goes, you can't see the forest for the trees.

Before going too far in that direction, it's a good idea to stop dissecting out elements and to start looking for the *interconnections*, the relationships that hold the elements together.

The interconnections in the tree system are the physical flows and chemical reactions that govern the tree's metabolic processes—the signals that allow one part to respond to what is happening in another part. For example, as the leaves lose water on a sunny day, a drop in pressure in the water-carrying vessels allows the roots to take in more water. Conversely, if the roots experience dry soil, the loss of water pressure signals the leaves to close their pores, so as not to lose even more precious water.

As the days get shorter in the temperate zones, a deciduous tree puts forth chemical messages that cause nutrients to migrate out of the leaves into the trunk and roots and that weaken the stems, allowing the leaves to fall. There even seem to be messages that cause some trees to make repellent chemicals or harder cell walls if just one part of the plant is attacked by insects. No one understands all the relationships that allow a tree to do what it does. That lack of knowledge is not surprising. It's easier to learn about a system's elements than about its interconnections.

In the university system, interconnections include the standards for admission, the requirements for degrees, the examinations and grades, the budgets and money flows, the gossip, and most important, the communication of knowledge that is, presumably, the purpose of the whole system.

Some interconnections in systems are actual physical flows, such as the water in the tree's trunk or the students progressing through a university. Many interconnections are flows of information—signals that go to decision points or action points within a system. These kinds of interconnections are often harder to see, but the system reveals them to those who look. Students may use informal information about the probability of getting a good grade to decide what courses to take. A consumer decides what to buy using information about his or her income, savings, credit rating, stock of goods at home, prices, and availability of goods for purchase. Governments need information about kinds and quantities of water pollution before they can create sensible regulations to reduce that pollution. (Note that information about the existence of a problem may be necessary but not sufficient to trigger action—information about resources, incentives, and consequences is necessary too.)

Many of the interconnections in systems operate through the flow of information. Information holds systems together and plays a great role in determining how they operate.

If information-based relationships are hard to see, *functions* or *purposes* are even harder. A system's function or purpose is not necessarily spoken, written, or expressed explicitly, except through the operation of the system. The best way to deduce the system's purpose is to watch for a while to see how the system behaves.

If a frog turns right and catches a fly, and then turns left and catches a fly, and then turns around backward and catches a fly, the purpose of the frog has to do not with turning left or right or backward but with catching flies. If a government proclaims its interest in protecting the environment but allocates little money or effort toward that goal, environmental protection is not, in fact, the government's purpose. Purposes are deduced from behavior, not from rhetoric or stated goals.

A NOTE ON LANGUAGE

The word *function* is generally used for a nonhuman system, the word *purpose* for a human one, but the distinction is not absolute, since so many systems have both human and nonhuman elements.

The function of a thermostat-furnace system is to keep a building at a given temperature. One function of a plant is to bear seeds and create more plants. One purpose of a national economy is, judging from its behavior, to keep growing larger. An important function of almost every system is to ensure its own perpetuation.

System purposes need not be human purposes and are not necessarily those intended by any single actor within the system. In fact, one of the most frustrating aspects of systems is that the purposes of subunits may add up to an overall behavior that no one wants. No one intends to produce a society with rampant drug addiction and crime, but consider the combined purposes and consequent actions of the actors involved:

- desperate people who want quick relief from psychological pain
- farmers, dealers, and bankers who want to earn money
- pushers who are less bound by civil law than are the police who oppose them
- governments that make harmful substances illegal and use police power to interdict them
- wealthy people living in close proximity to poor people
- nonaddicts who are more interested in protecting themselves than in encouraging recovery of addicts

Altogether, these make up a system from which it is extremely difficult to eradicate drug addiction and crime.

Systems can be nested within systems. Therefore, there can be purposes within purposes. The purpose of a university is to discover and preserve knowledge and pass it on to new generations. Within the university, the purpose of a student may be to get good grades, the purpose of a professor may be to get tenure, the purpose of an administrator may be to balance the budget. Any of those sub-purposes could come into conflict with the overall purpose—the student could cheat, the professor could ignore the students in order to publish papers, the administrator could balance the budget by firing professors. Keeping sub-purposes and overall system purposes in harmony is an essential function of successful systems. I'll get back to this point later when we come to hierarchies.

You can understand the relative importance of a system's elements, interconnections, and purposes by imagining them changed one by one. Changing elements usually has the least effect on the system. If you change all the players on a football team, it is still recognizably a football team. (It may play much better or much worse—particular elements in a system can indeed be important.) A tree changes its cells constantly, its leaves every year or so, but it is still essentially the same tree. Your body replaces most of its cells every few weeks, but it goes on being your body. The university has a constant flow of students and a slower flow of professors and administrators, but it is still a university. In fact it is still the same university, distinct in subtle ways from others, just as General Motors and the U.S. Congress somehow maintain their identities even though all their members change. A system generally goes on being itself, changing only slowly if at all, even with complete substitutions of its elements—as long as its interconnections and purposes remain intact.

> The least obvious part of the system, its function or purpose, is often the most crucial determinant of the system's behavior.

If the interconnections change, the system may be greatly altered. It may even become unrecognizable, even though the same players are on the team. Change the rules from those of football to those of basketball, and you've got, as they say, a whole new ball game. If you change the interconnections in the tree—say that instead of taking in carbon dioxide and emitting oxygen, it does the reverse—it would no longer be a tree. (It would be an animal.) If in a university the students graded the professors, or if arguments were won by force instead of reason, the place would need a different name. It might be an interesting organization, but it would not be a university. Changing interconnections in a system can change it dramatically.

Changes in function or purpose also can be drastic. What if you keep the players and the rules but change the purpose—from winning to losing, for example? What if the function of a tree were not to survive and reproduce but to capture all the nutrients in the soil and grow to unlimited size? People have imagined many purposes for a university besides disseminating knowledge—making money, indoctrinating people, winning football games. A change in purpose changes a system profoundly, even if every element and interconnection remains the same.

To ask whether elements, interconnections, or purposes are most important in a system is to ask an unsystemic question. All are essential. All interact. All have their roles. But the least obvious part of the system, its function or purpose, is often the most crucial determinant of the system's behavior. Interconnections are also critically important. Changing relationships usually changes system behavior. The elements, the parts of systems we are most likely to notice, are often (not always) least important in defining the unique characteristics of the system—*unless changing an element also results in changing relationships or purpose.*

Changing just one leader at the top—from a Brezhnev to a Gorbachev, or from a Carter to a Reagan—may or may not turn an entire nation in a new direction, though its land, factories, and hundreds of millions of people remain exactly the same. A leader can make that land and those factories and people play a different game with new rules, or can direct the play toward a new purpose.

And conversely, because land, factories, and people are long-lived, slowly changing, physical elements of the system, there is a limit to the rate at which any leader can turn the direction of a nation.

Bathtubs 101—Understanding System Behavior over Time

> Information contained in nature . . . allows us a partial reconstruction of the past. . . . The development of the meanders in a river, the increasing complexity of the earth's crust . . . are information-storing devices in the same manner that genetic systems are. . . . Storing information means increasing the complexity of the mechanism.
>
> —Ramon Margalef[2]

A **stock** is the foundation of any system. Stocks are the elements of the system that you can see, feel, count, or measure at any given time. A system stock is just what it sounds like: a store, a quantity, an accumulation of material or information that has built up over time. It may be the water in a bathtub, a population, the books in a bookstore, the wood in a tree, the money in a bank, your own self-confidence. A stock does not have to be physical. Your reserve of good will toward others or your supply of hope that the world can be better are both stocks.

A stock is the memory of the history of changing flows within the system.

Stocks change over time through the actions of a **flow**. Flows are filling and draining, births and deaths, purchases and sales, growth and decay, deposits and withdrawals, successes and failures. A stock, then, is the present memory of the history of changing flows within the system.

Figure 1. How to read stock-and-flow diagrams. In this book, stocks are shown as boxes, and flows as arrow-headed "pipes" leading into or out of the stocks. The small T on each flow signifies a "faucet;" it can be turned higher or lower, on or off. The "clouds" stand for wherever the flows come from and go to—the sources and sinks that are being ignored for the purposes of the present discussion.

For example, an underground mineral deposit is a stock, out of which comes a flow of ore through mining. The inflow of ore into a mineral deposit is minute in any time period less than eons. So I have chosen to draw (Figure 2) a simplified picture of the system without any inflow. *All* system diagrams and descriptions are simplified versions of the real world.

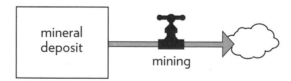

Figure 2. A stock of minerals depleted by mining.

Water in a reservoir behind a dam is a stock, into which flow rain and river water, and out of which flows evaporation from the reservoir's surface as well as the water discharged through the dam.

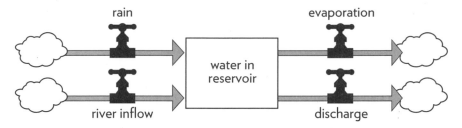

Figure 3. A stock of water in a reservoir with multiple inflows and outflows.

The volume of wood in the living trees in a forest is a stock. Its inflow is the growth of the trees. Its outflows are the natural deaths of trees and the harvest by loggers. The logging harvest flows into another stock, perhaps an inventory of lumber at a mill. Wood flows out of the inventory stock as lumber sold to customers.

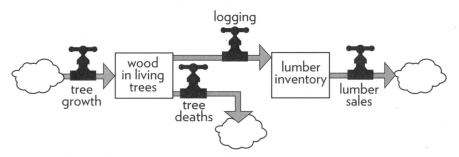

Figure 4. A stock of lumber linked to a stock of trees in a forest.

If you understand the **dynamics** of stocks and flows—their behavior over time—you understand a good deal about the behavior of complex systems. And if you have had much experience with a bathtub, you understand the dynamics of stocks and flows.

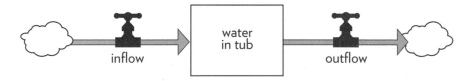

Figure 5. The structure of a bathtub system—one stock with one inflow and one outflow.

Imagine a bathtub filled with water, with its drain plugged up and its faucets turned off—an unchanging, undynamic, boring system. Now mentally pull the plug. The water runs out, of course. The level of water in the tub goes down until the tub is empty.

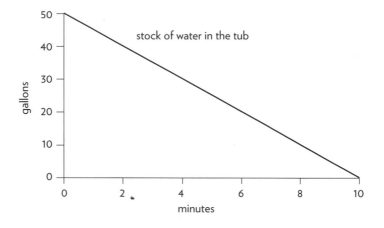

Figure 6. Water level in a tub when the plug is pulled.

A NOTE ON READING GRAPHS OF BEHAVIOR OVER TIME

Systems thinkers use graphs of system behavior to understand trends over time, rather than focusing attention on individual events. We also use behavior-over-time graphs to learn whether the system is approaching a goal or a limit, and if so, how quickly.

The variable on the graph may be a stock or a flow. The pattern—the shape of the variable line—is important, as are the points at which that line changes shape or direction. The precise numbers on the axes are often less important.

The horizontal axis of time allows you to ask questions about what came before, and what might happen next. It can help you focus on the time horizon appropriate to the question or problem you are investigating.

Now imagine starting again with a full tub, and again open the drain, but this time, when the tub is about half empty, turn on the inflow faucet so the rate of water flowing in is just equal to that flowing out. What happens?

The amount of water in the tub stays constant at whatever level it had reached when the inflow became equal to the outflow. It is in a state of **dynamic equilibrium**—its level does not change, although water is continuously flowing through it.

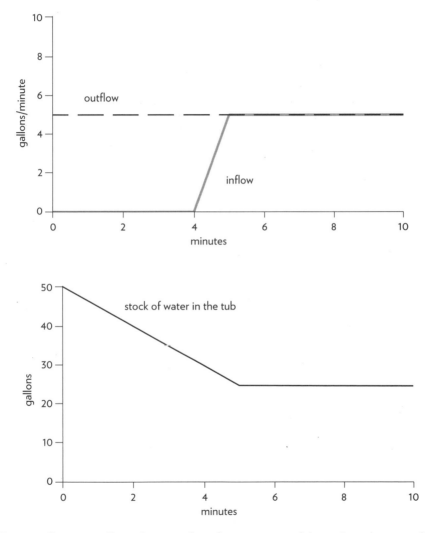

Figure 7. Constant outflow, inflow turned on after 5 minutes, and the resulting changes in the stock of water in the tub.

Imagine turning the inflow on somewhat harder while keeping the outflow constant. The level of water in the tub slowly rises. If you then turn the inflow faucet down again to match the outflow exactly, the water in the tub will stop rising. Turn it down some more, and the water level will fall slowly.

This model of a bathtub is a very simple system with just one stock, one inflow, and one outflow. Over the time period of interest (minutes), I have assumed that evaporation from the tub is insignificant, so I have not included that outflow. All models, whether mental models or mathematical models, are simplifications of the real world. You know all the dynamic possibilities of this bathtub. From it you can deduce several important principles that extend to more complicated systems:

- As long as the sum of all inflows exceeds the sum of all outflows, the level of the stock will rise.
- As long as the sum of all outflows exceeds the sum of all inflows, the level of the stock will fall.
- If the sum of all outflows equals the sum of all inflows, the stock level will not change; it will be held in dynamic equilibrium at whatever level it happened to be when the two sets of flows became equal.

The human mind seems to focus more easily on stocks than on flows. On top of that, when we do focus on flows, we tend to focus on *inflows* more easily than on outflows. Therefore, we sometimes miss seeing that we can fill a bathtub not only by increasing the inflow rate, but also by decreasing the outflow rate. Everyone understands that you can prolong the life of an oil-based economy by discovering new oil deposits. It seems to be harder to understand that the same result can be achieved by burning less oil. A breakthrough in energy efficiency is equivalent, in its effect on the stock of available oil, to the discovery of a new oil field—although different people profit from it.

A stock can be increased by decreasing its outflow rate as well as by increasing its inflow rate. There's more than one way to fill a bathtub!

Similarly, a company can build up a larger workforce by more hiring, or it can do the same thing by reducing the rates of quitting and firing. These two strategies may have very different costs. The wealth of a nation can be boosted by investment to build up a larger stock of factories and machines. It also can be boosted, often more cheaply, by decreasing the rate at which factories and machines wear out, break down, or are discarded.

You can adjust the drain or faucet of a bathtub—the flows—abruptly, but it is much more difficult to change the level of water—the stock—quickly. Water can't run out the drain instantly, even if you open the drain all the way. The tub can't fill up immediately, even with the inflow faucet on full blast. *A stock takes time to change, because flows take time to flow.* That's a vital point, a key to understanding why systems behave as they do. Stocks usually change slowly. They can act as delays, lags, buffers, ballast, and sources of momentum in a system. Stocks, especially large ones, respond to change, even sudden change, only by gradual filling or emptying.

> Stocks generally change slowly, even when the flows into or out of them change suddenly. Therefore, **stocks act as delays or buffers or shock absorbers in systems.**

People often underestimate the inherent momentum of a stock. It takes a long time for populations to grow or stop growing, for wood to accumulate in a forest, for a reservoir to fill up, for a mine to be depleted. An economy cannot build up a large stock of functioning factories and highways and electric plants overnight, even if a lot of money is available. Once an economy has a lot of oil-burning furnaces and automobile engines, it cannot change quickly to furnaces and engines that burn a different fuel, even if the price of oil suddenly changes. It has taken decades to accumulate the stratospheric pollutants that destroy the earth's ozone layer; it will take decades for those pollutants to be removed.

Changes in stocks set the pace of the dynamics of systems. Industrialization cannot proceed faster than the rate at which factories and machines can be constructed and the rate at which human beings can be educated to run and maintain them. Forests can't grow overnight. Once contaminants have accumulated in groundwater, they can be washed out only at the rate of groundwater turnover, which may take decades or even centuries.

The time lags that come from slowly changing stocks can cause problems in systems, but they also can be sources of stability. Soil that has accumulated over centuries rarely erodes all at once. A population that has learned many skills doesn't forget them immediately. You can pump groundwater faster than the rate it recharges for a long time before the aquifer is drawn down far enough to be damaged. The time lags imposed by stocks allow room to maneuver, to experiment, and to revise policies that aren't working.

If you have a sense of the rates of change of stocks, you don't expect things to happen faster than they can happen. You don't give up too soon.

You can use the opportunities presented by a system's momentum to guide it toward a good outcome—much as a judo expert uses the momentum of an opponent to achieve his or her own goals.

There is one more important principle about the role of stocks in systems, a principle that will lead us directly to the concept of feedback. The presence of stocks allows inflows and outflows to be independent of each other and temporarily out of balance with each other.

> **Stocks allow inflows and outflows to be decoupled and to be independent and temporarily out of balance with each other.**

It would be hard to run an oil company if gasoline had to be produced at the refinery at exactly the rate the cars were burning it. It isn't feasible to harvest a forest at the precise rate at which the trees are growing. Gasoline in storage tanks and wood in the forest are both stocks that permit life to proceed with some certainty, continuity, and predictability, even though flows vary in the short term.

Human beings have invented hundreds of stock-maintaining mechanisms to make inflows and outflows independent and stable. Reservoirs enable residents and farmers downriver to live without constantly adjusting their lives and work to a river's varying flow, especially its droughts and floods. Banks enable you temporarily to earn money at a rate different from how you spend. Inventories of products along a chain from distributors to wholesalers to retailers allow production to proceed smoothly although customer demand varies, and allow customer demand to be filled even though production rates vary.

Most individual and institutional decisions are designed to regulate the levels in stocks. If inventories rise too high, then prices are cut or advertising budgets are increased, so that sales will go up and inventories will fall. If the stock of food in your kitchen gets low, you go to the store. As the stock of growing grain rises or fails to rise in the fields, farmers decide whether to apply water or pesticide, grain companies decide how many barges to book for the harvest, speculators bid on future values of the harvest, cattle growers build up or cut down their herds. Water levels in reservoirs cause all sorts of corrective actions if they rise too high or fall too low. The same can be said for the stock of money in your wallet, the oil reserves owned by an oil company, the pile of woodchips feeding a paper mill, and the concentration of pollutants in a lake.

People monitor stocks constantly and make decisions and take actions designed to raise or lower stocks or to keep them within acceptable ranges. Those decisions add up to the ebbs and flows, successes and problems, of all sorts of systems. Systems thinkers see the world as a collection of stocks along with the mechanisms for regulating the levels in the stocks by manipulating flows.

That means system thinkers see the world as a collection of "feedback processes."

How the System Runs Itself—Feedback

> Systems of information-feedback control are fundamental to all life and human endeavor, from the slow pace of biological evolu- tion to the launching of the latest space satellite. . . . Everything we do as individuals, as an industry, or as a society is done in the context of an information-feedback system.
>
> —Jay W. Forrester[3]

When a stock grows by leaps and bounds or declines swiftly or is held within a certain range no matter what else is going on around it, it is likely that there is a control mechanism at work. In other words, if you see a behavior that persists over time, there is likely a mechanism creating that consistent behavior. That mechanism operates through a **feedback loop**. It is the consistent behavior pattern over a long period of time that is the first hint of the existence of a feedback loop.

A feedback loop is formed when changes in a stock affect the flows into or out of that same stock. A feedback loop can be quite simple and direct. Think of an interest-bearing savings account in a bank. The total amount of money in the account (the stock) affects how much money comes into the account as interest. That is because the bank has a rule that the account earns a certain percent interest each year. The total dollars of interest paid into the account each year (the flow in) is not a fixed amount, but varies with the size of the total in the account.

You experience another fairly direct kind of feedback loop when you get your bank statement for your checking account each month. As your level of available cash in the checking account (a stock) goes down, you may decide to work more hours and earn more money. The money entering your bank account is a flow that you can adjust in order to increase your stock of cash to a more desirable level. If your bank account then grows very large, you may feel free to work less (decreasing the inflow). This kind of feedback loop is keeping your level of cash available within a range that is acceptable to you. You can see that adjusting your earnings is not the only feedback loop that works on your stock of cash. You also may be able to adjust the outflow of money from your account, for example. You can imagine an outflow-adjusting feedback loop for spending.

Feedback loops can cause stocks to maintain their level within a range or grow or decline. In any case, the flows into or out of the stock are adjusted because of changes in the size of the stock itself. Whoever or whatever is monitoring the stock's level begins a corrective process, adjusting rates of inflow or outflow (or both) and so changing the stock's level. The stock level feeds back through a chain of signals and actions to control itself.

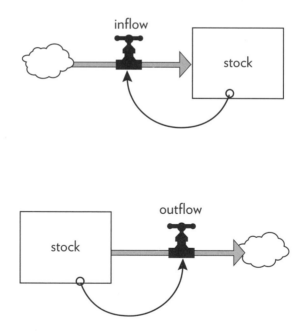

Figure 8. How to read a stock-and-flow diagram with feedback loops. Each diagram distinguishes the stock, the flow that changes the stock, and the information link (shown as a thin, curved arrow) that directs the action. It emphasizes that action or change always proceeds through adjusting flows.

Not all systems have feedback loops. Some systems are relatively simple open-ended chains of stocks and flows. The chain may be affected by outside factors, but the levels of the chain's stocks don't affect its flows. However, those systems that contain feedback loops are common and may be quite elegant or rather surprising, as we shall see.

A feedback loop is a closed chain of causal connections from a stock, through a set of decisions or rules or physical laws or actions that are dependent on the level of the stock, and back again through a flow to change the stock.

Stabilizing Loops—Balancing Feedback

One common kind of feedback loop stabilizes the stock level, as in the checking-account example. The stock level may not remain completely fixed, but it does stay within an acceptable range. What follows are some more stabilizing feedback loops that may be familiar to you. These examples start to detail some of the steps within a feedback loop.

If you're a coffee drinker, when you feel your energy level run low, you may grab a cup of hot black stuff to perk you up again. You, as the coffee drinker, hold in your mind a desired stock level (energy for work). The purpose of this caffeine-delivery system is to keep your actual stock level near or at your desired level. (You may have other purposes for drinking coffee as well: enjoying the flavor or engaging in a social activity.) It is the

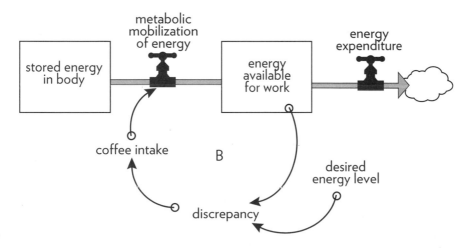

Figure 9. Energy level of a coffee drinker.

gap, the discrepancy, between your actual and desired levels of energy for work that drives your decisions to adjust your daily caffeine intake.

Notice that the labels in Figure 9, like all the diagram labels in this book, are direction-free. The label says "stored energy in body" not "*low* energy level," "coffee intake" not "*more* coffee." That's because feedback loops often can operate in two directions. In this case, the feedback loop can correct an oversupply as well as an undersupply. If you drink too much coffee and find yourself bouncing around with extra energy, you'll lay off the caffeine for a while. High energy creates a discrepancy that says "too much," which then causes you to reduce your coffee intake until your energy level settles down. The diagram is intended to show that the loop works to drive the stock of energy in either direction.

I could have shown the inflow of energy coming from a cloud, but instead I made the system diagram slightly more complicated. *Remember—all system diagrams are simplifications of the real world.* We each choose how much complexity to look at. In this example, I drew another stock—the stored energy in the body that can be activated by the caffeine. I did that to indicate that there is more to the system than one simple loop. As every coffee drinker knows, caffeine is only a short-term stimulant. It lets you run your motor faster, but it doesn't refill your personal fuel tank. Eventually the caffeine high wears off, leaving the body more energy-deficient than it was before. That drop could reactivate the feedback loop and generate another trip to the coffee pot. (See the discussion of addiction later in this book.) Or it could activate some longer-term and healthier feedback responses: Eat some food, take a walk, get some sleep.

This kind of stabilizing, goal-seeking, regulating loop is called a **balancing feedback loop**, so I put a B inside the loop in the diagram. Balancing feedback loops are *goal-seeking* or *stability-seeking*. Each tries to keep a stock at a given value or within a range of values. A balancing feedback loop opposes whatever direction of change is imposed on the system. If you push a stock too far up, a balancing loop will try to pull it back down. If you shove it too far down, a balancing loop will try to bring it back up.

Here's another balancing feedback loop that involves coffee, but one that works through physical law rather than human decision. A hot cup of coffee will gradually cool down to room temperature. Its rate of cooling depends on the difference between the temperature of the coffee and the temperature of the room. The greater the difference, the faster the coffee will cool. The loop works the other way too—if you make iced coffee on a hot day, it will warm up until it has the same temperature as the room. The function of this system is to bring the discrepancy between coffee's temperature and room's temperature to zero, no matter what the direction of the discrepancy.

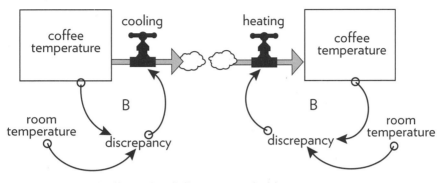

Figure 10. A cup of coffee cooling (*left*) or warming (*right*).

Starting with coffee at different temperatures, from just below boiling to just above freezing, the graphs in Figure 11 show what will happen to the temperature over time (if you don't drink the coffee). You can see here the "homing" behavior of a balancing feedback loop. Whatever the initial value of the system stock (coffee temperature in this case), whether it is above or below the "goal" (room temperature), the feedback loop brings it toward

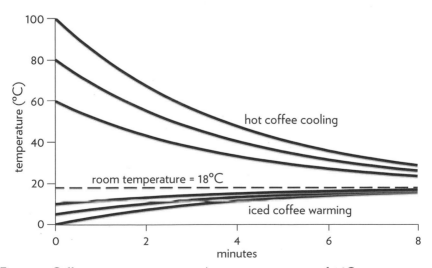

Figure 11. Coffee temperature as it approaches a room temperature of 18°C.

the goal. The change is faster at first, and then slower, as the discrepancy between the stock and the goal decreases.

Balancing feedback loops are equilibrating or goal-seeking structures in systems and are both sources of stability and sources of resistance to change.

This behavior pattern—gradual approach to a system-defined goal— also can be seen when a radioactive element decays, when a missile finds its target, when an asset depreciates, when a reservoir is brought up or down to its desired level, when your body adjusts its blood-sugar concentration, when you pull your car to a stop at a stoplight. You can think of many more examples. The world is full of goal-seeking feedback loops.

The presence of a feedback mechanism doesn't necessarily mean that the mechanism works *well*. The feedback mechanism may not be strong enough to bring the stock to the desired level. Feedbacks—the interconnections, the information part of the system—can fail for many reasons. Information can arrive too late or at the wrong place. It can be unclear or incomplete or hard to interpret. The action it triggers may be too weak or delayed or resource-constrained or simply ineffective. The goal of the feedback loop may never be reached by the actual stock. But in the simple example of a cup of coffee, the drink eventually will reach room temperature.

Runaway Loops—Reinforcing Feedback

I'd need rest to refresh my brain, and to get rest it's necessary to travel, and to travel one must have money, and in order to get money you have to work. . . . I am in a vicious circle . . . from which it is impossible to escape.

— Honoré Balzac,[4] 19th century novelist and playwright

Here we meet a very important feature. It would seem as if this were circular reasoning; profits fell because investment fell, and investment fell because profits fell.

— Jan Tinbergen,[5] economist

The second kind of feedback loop is amplifying, reinforcing, self-multiplying, snowballing—a vicious or virtuous circle that can cause healthy growth or runaway destruction. It is called a **reinforcing feedback loop**, and will be noted with an R in the diagrams. It generates more input to a stock the more that is already there (and less input the less that is already there). A reinforcing feedback loop enhances whatever direction of change is imposed on it.

For example:

- When we were kids, the more my brother pushed me, the more I pushed him back, so the more he pushed me back, so the more I pushed him back.
- The more prices go up, the more wages have to go up if people are to maintain their standards of living. The more wages go up, the more prices have to go up to maintain profits. This means that wages have to go up again, so prices go up again.
- The more rabbits there are, the more rabbit parents there are to make baby rabbits. The more baby rabbits there are, the more grow up to become rabbit parents, to have even more baby rabbits.
- The more soil is eroded from the land, the less plants are able to grow, so the fewer roots there are to hold the soil, so the more soil is eroded, so less plants can grow.
- The more I practice piano, the more pleasure I get from the sound, and so the more I play the piano, which gives me more practice.

Reinforcing loops are found wherever a system element has the ability to reproduce itself or to grow as a constant fraction of itself. Those elements include populations and economies. Remember the example of

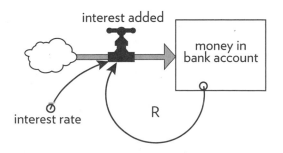

Figure 12. Interest-bearing bank account.

the interest-bearing bank account? The more money you have in the bank, the more interest you earn, which is added to the money already in the bank, where it earns even more interest.

Figure 13 shows how this reinforcing loop multiplies money, starting with $100 in the bank, and assuming no deposits and no withdrawals over a period of twelve years. The five lines show five different interest rates, from 2 percent to 10 percent per year.

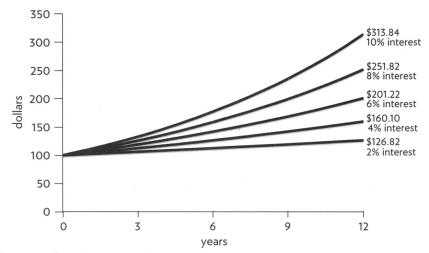

Figure 13. Growth in savings with various interest rates.

This is not simple linear growth. It is not constant over time. The growth of the bank account at lower interest rates may look linear in the first few years. But, in fact, growth goes faster and faster. The more is there, the more is added. This kind of growth is called "exponential." It's either good news or bad news, depending on what is growing—money in the bank, people with HIV/AIDS, pests in a cornfield, a national economy, or weapons in an arms race.

Reinforcing feedback loops are self-enhancing, leading to exponential growth or to runaway collapses over time. They are found whenever a stock has the capacity to reinforce or reproduce itself.

In Figure 14, the more machines and factories (collectively called "capital") you have, the more goods and services ("output") you can produce. The more output you can produce, the more you can invest in new machines and factories. The more you make, the more capacity you have to make even more. This reinforcing feedback loop is the central engine of growth in an economy.

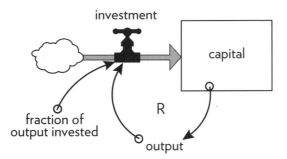

Figure 14. Reinvestment in capital.

By now you may be seeing how basic balancing and reinforcing feedback loops are to systems. Sometimes I challenge my students to try to think of any human decision that occurs *without* a feedback loop—that is, a decision that is made without regard to any information about the level of the stock it influences. Try thinking about that yourself. The more you do, the more you'll begin to see feedback loops everywhere.

The most common "non-feedback" decisions students suggest are falling in love and committing suicide. I'll leave it to you to decide whether you think these are *actually* decisions made with no feedback involved.

Watch out! If you see feedback loops everywhere, you're already in danger of becoming a systems thinker! Instead of seeing only how A causes B, you'll begin to wonder how B may *also* influence A—and how A might reinforce or reverse itself. When you hear in the nightly news that the Federal Reserve

HINT ON REINFORCING LOOPS AND DOUBLING TIME

Because we bump into reinforcing loops so often, it is handy to know this shortcut: The time it takes for an exponentially growing stock to double in size, the "doubling time," equals approximately 70 divided by the growth rate (expressed as a percentage).

Example: If you put $100 in the bank at 7% interest per year, you will double your money in 10 years (70 ÷ 7 = 10). If you get only 5% interest, your money will take 14 years to double.

Bank has done something to control the economy, you'll also see that the economy must have done something to affect the Federal Reserve Bank. When someone tells you that population growth causes poverty, you'll ask yourself how poverty may cause population growth.

THINK ABOUT THIS:

If A causes B, is it possible that B also causes A?

You'll be thinking not in terms of a static world, but a dynamic one. You'll stop looking for who's to blame; instead you'll start asking, "What's the system?" The concept of feedback opens up the idea that a system can cause its own behavior.

So far, I have limited this discussion to one kind of feedback loop at a time. Of course, in real systems feedback loops rarely come singly. They are linked together, often in fantastically complex patterns. A single stock is likely to have several reinforcing and balancing loops of differing strengths pulling it in several directions. A single flow may be adjusted by the contents of three or five or twenty stocks. It may fill one stock while it drains another and feeds into decisions that alter yet another. The many feedback loops in a system tug against each other, trying to make stocks grow, die off, or come into balance with each other. As a result, complex systems do much more than stay steady or explode exponentially or approach goals smoothly—as we shall see.

TWO

A Brief Visit to the Systems Zoo

The . . . goal of all theory is to make the . . . basic elements as simple and as few as possible without having to surrender the adequate representation of . . . experience.

—Albert Einstein,[1] physicist

One good way to learn something new is through specific examples rather than abstractions and generalities, so here are several common, simple but important examples of systems that are useful to understand in their own right and that will illustrate many general principles of complex systems.

This collection has some of the same strengths and weaknesses as a zoo.[2] It gives you an idea of the large variety of systems that exist in the world, but it is far from a complete representation of that variety. It groups the animals by family—monkeys here, bears there (single-stock systems here, two-stock systems there)—so you can observe the characteristic behaviors of monkeys, as opposed to bears. But, like a zoo, this collection is too neat. To make the animals visible and understandable, it separates them from each other and from their normal concealing environment. Just as zoo animals more naturally occur mixed together in ecosystems, so the systems animals described here normally connect and interact with each other and with others not illustrated here—all making up the buzzing, hooting, chirping, changing complexity in which we live.

Ecosystems come later. For the moment, let's look at one system animal at a time.

One-Stock Systems

A Stock with Two Competing Balancing Loops—a Thermostat

You already have seen the "homing in" behavior of the goal-seeking balancing feedback loop—the coffee cup cooling. What happens if there are two such loops, trying to drag a single stock toward two different goals?

One example of such a system is the thermostat mechanism that regulates the heating of your room (or cooling, if it is connected to an air conditioner instead of a furnace). Like all models, the representation of a thermostat in Figure 15 is a simplification of a real home heating system.

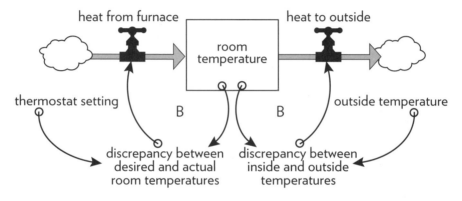

Figure 15. Room temperature regulated by a thermostat and furnace.

Whenever the room temperature falls below the thermostat setting, the thermostat detects a discrepancy and sends a signal that turns on the heat flow from the furnace, warming the room. When the room temperature rises again, the thermostat turns off the heat flow. This straightforward, stock-maintaining, balancing feedback loop is shown on the left side of Figure 15. If there were nothing else in the system, and if you start with a cold room with the thermostat set at 18°C (65°F), it would behave as shown in Figure 16. The furnace comes on, and the room warms up. When the room temperature reaches the thermostat setting, the furnace goes off, and the room stays right at the target temperature.

However, this is not the only loop in the system. Heat also leaks to the outside. The outflow of heat is governed by the second balancing feedback loop, shown on the right side of Figure 15. It is always trying to make the room temperature equal to the outside, just like a coffee cup cooling. If

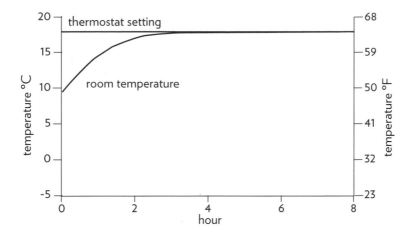

Figure 16. A cold room warms quickly to the thermostat setting.

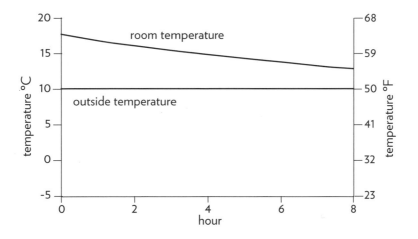

Figure 17. A warm room cools very slowly to the outside temperature of 10°C.

this were the only loop in the system (if there were no furnace), Figure 17 shows what would happen, starting with a warm room on a cold day.

The assumption is that room insulation is not perfect, and so some heat leaks out of the warm room to the cool outdoors. The better the insulation, the slower the drop in temperature would be.

Now, what happens when these two loops operate at the same time? Assuming that there is sufficient insulation and a properly sized furnace, the heating loop dominates the cooling loop. You end up with a warm room (see Figure 18), even starting with a cold room on a cold day.

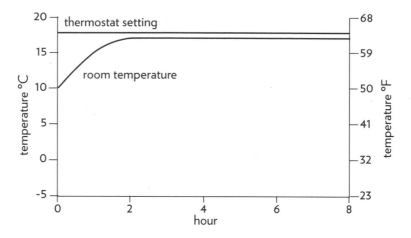

Figure 18. The furnace warms a cool room, even as heat continues to leak from the room.

As the room heats up, the heat flowing out of it increases, because there's a larger gap between inside and outside temperatures. But the furnace keeps putting in more heat than the amount that leaks out, so the room warms nearly to the target temperature. At that point, the furnace cycles off and on as it compensates for the heat constantly flowing out of the room.

The thermostat is set at 18°C (65°F) in this simulation, but the room temperature levels off slightly below 18°C (65°F). That's because of the leak to the outside, which is draining away some heat even as the furnace is getting the signal to put it back. This is a characteristic and sometimes surprising behavior of a system with competing balancing loops. It's like trying to keep a bucket full when there's a hole in the bottom. To make things worse, water leaking out of the hole is governed by a feedback loop; the more water in the bucket, the more the water pressure at the hole increases, so the flow out increases! In this case, we are trying to keep the room warmer than the outside and the warmer the room is, the faster it loses heat to the outside. It takes time for the furnace to correct for the increased heat loss—and in that minute still more heat leaks out. In a well-insulated house, the leak will be slower and so the house more comfortable than a poorly insulated one, even a poorly insulated house with a big furnace.

With home heating systems, people have learned to set the thermostat slightly higher than the actual temperature they are aiming at. Exactly how much higher can be a tricky question, because the outflow rate is higher on cold days than on warm days. But for thermostats this control problem isn't serious. You can muddle your way to a thermostat setting you can live with.

For other systems with this same structure of competing balancing loops, the fact that the stock goes on changing while you're trying to control it can create real problems. For example, suppose you're trying to maintain a store inventory at a certain level. You can't instantly order new stock to make up an immediately apparent shortfall. If you don't account for the goods that will be sold while you are waiting for the order to come in, your inventory will never be quite high enough. You can be fooled in the same way if you're trying to maintain a cash balance at a certain level, or the level of water in a reservoir, or the concentration of a chemical in a continuously flowing reaction system.

There's an important general principle here, and also one specific to the thermostat structure. First the general one: The information delivered by a feedback loop can only affect future behavior; it can't deliver the information, and so can't have an impact fast enough to correct behavior that drove the current feedback. A person in the system who makes a decision based on the feedback can't change the behavior of the system that drove the current feedback; the decisions he or she makes will affect only future behavior.

Why is that important? Because it means there will always be delays in responding. It says that a flow can't react instantly to a flow. It can react only to a change in a stock, and only after a slight delay to register the incoming information. In the bathtub, it takes a split second of time to assess the depth of the water and decide to adjust the flows. Many economic models make a mistake in this matter by assuming that consumption or production can respond immediately, say, to a change in price. That's one of the reasons why real economies tend not to behave exactly like many economic models.

> The information delivered by a feedback loop—even nonphysical feedback—can only affect future behavior; it can't deliver a signal fast enough to correct behavior that drove the current feedback. Even nonphysical information takes time to feedback into the system.

The specific principle you can deduce from this simple system is that you must remember in thermostat-like systems to take into account whatever draining or filling processes are going on. If you don't, you won't achieve the target level of your stock. If you want your room temperature to be at 18°C (65°F), you have to set the thermostat a little above the desired temperature. If you want to pay off your credit card (or the national debt), you have to raise your repayment rate high enough to cover the charges you incur while you're paying (including interest). If you're gearing up your work force to a higher level, you have to hire fast enough to correct for those who quit while you are hiring. In other words, your mental model of the system needs to include all the important flows, or you will be surprised by the system's behavior.

> A stock-maintaining balancing feedback loop must have its goal set appropriately to compensate for draining or inflowing processes that affect that stock. Otherwise, the feedback process will fall short of or exceed the target for the stock.

Before we leave the thermostat, we should see how it behaves with a varying outside temperature. Figure 19 shows a twenty-four-hour period of normal operation of a well-functioning thermostat system, with the outside temperature dipping well below freezing. The inflow of heat from the furnace nicely tracks the outflow of heat to the outside. The temperature in the room varies hardly at all once the room has warmed up.

Every balancing feedback loop has its breakdown point, where other loops pull the stock away from its goal more strongly than it can pull back. That can happen in this simulated thermostat system, if I weaken the power of the heating loop (a smaller furnace that cannot put out as much heat), or if I strengthen the power of the cooling loop (colder outside tempera-

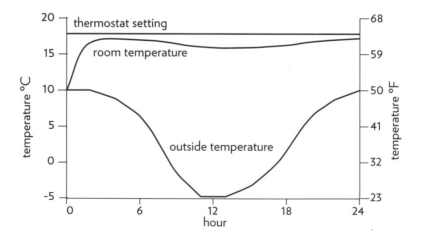

Figure 19. The furnace warms a cool room, even as heat leaks from the room and outside temperatures drop below freezing.

ture, less insulation, or larger leaks). Figure 20 shows what happens with the same outside temperatures as in Figure 19, but with faster heat loss from the room. At very cold temperatures, the furnace just can't keep up with the heat drain. The loop that is trying to bring the room temperature down to the outside temperature dominates the system for a while. The room gets pretty uncomfortable!

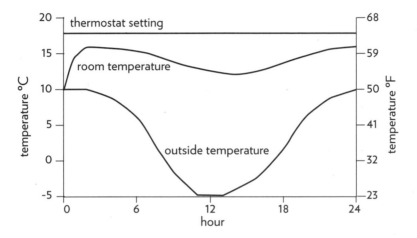

Figure 20. On a cold day, the furnace can't keep the room warm in this leaky house!

See if you can follow, as time unfolds, how the variables in Figure 20 relate to one another. At first, both the room and the outside temperatures are cool. The inflow of heat from the furnace exceeds the leak to the outside, and the room warms up. For an hour or two, the outside is mild enough that the furnace replaces most of the heat that's lost to the outside, and the room temperature stays near the desired temperature.

But as the outside temperature falls and the heat leak increases, the furnace cannot replace the heat fast enough. Because the furnace is generating less heat than is leaking out, the room temperature falls. Finally, the outside temperature rises again, the heat leak slows, and the furnace, still operating at full tilt, finally can pull ahead and start to warm the room again

Just as in the rules for the bathtub, whenever the furnace is putting in more heat than is leaking out, the room temperature rises. Whenever the inflow rate falls behind the outflow rate, the temperature falls. If you study the system changes on this graph for a while and relate them to the feedback-loop diagram of this system, you'll get a good sense of how the structural interconnections of this system—its two feedback loops and how they shift in strength relative to each other—lead to the unfolding of the system's behavior over time.

A Stock with One Reinforcing Loop and One Balancing Loop—Population and Industrial Economy

What happens when a reinforcing and a balancing loop are both pulling on the same stock? This is one of the most common and important system structures. Among other things, it describes every living population and every economy.

A population has a reinforcing loop causing it to grow through its birth rate, and a balancing loop causing it to die off through its death rate.

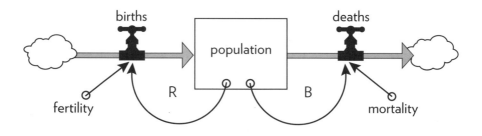

Figure 21. Population governed by a reinforcing loop of births and a balancing loop of deaths.

As long as fertility and mortality are constant (which in real systems they rarely are), this system has a simple behavior. It grows exponentially or dies off, depending on whether its reinforcing feedback loop determining births is stronger than its balancing feedback loop determining deaths.

For example, the 2007 world population of 6.6 billion people had a fertility rate of roughly 21 births a year for every 1,000 people in the population. Its mortality rate was 9 deaths a year out of every 1,000 people. Fertility was higher than mortality, so the reinforcing loop dominated the system. If those fertility and mortality rates continue unchanged, a child born now will see the world population more than double by the time he or she reaches the age of 60, as shown in Figure 22.

If, because of a terrible disease, the mortality rate were higher, say at 30

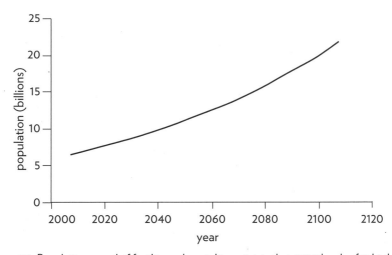

Figure 22. Population growth if fertility and mortality maintain their 2007 levels of 21 births and nine deaths per 1,000 people.

deaths per 1,000, while the fertility rate remained at 21, then the death loop would dominate the system. More people would die each year than would be born, and the population would gradually decrease (Figure 23).

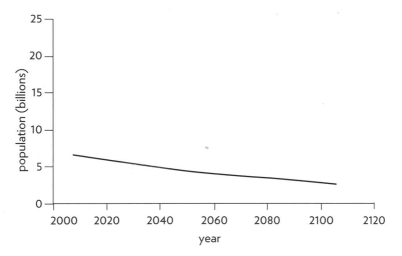

Figure 23. Population decline if fertility remains at 2007 level (21 births per 1,000) but mortality is much higher, 30 deaths per 1,000.

Things get more interesting when fertility and mortality change over time. When the United Nations makes long-range population projections, it generally assumes that, as countries become more developed, average fertility will decline (approaching replacement where on average each woman has 1.85 children). Until recently, assumptions about mortality were that it would also decline, but more slowly (because it is already low in most parts of the world). However, because of the epidemic of HIV/AIDS, the UN now assumes the trend of increasing life expectancy over the next fifty years will slow in regions affected by HIV/AIDS.

Changing flows (fertility and mortality) create a change in the behavior over time of the stock (population)—the line bends. If, for example, world fertility falls steadily to equal mortality by the year 2035 and they both stay constant thereafter, the population will level off, births exactly balancing deaths in dynamic equilibrium, as in Figure 24.

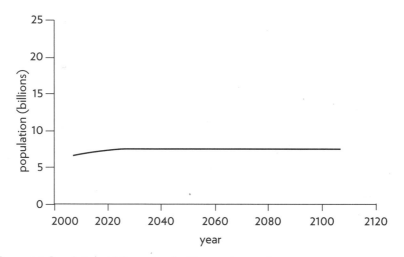

Figure 24. Population stabilizes when fertility equals mortality.

This behavior is an example of **shifting dominance** of feedback loops. Dominance is an important concept in systems thinking. When one loop dominates another, it has a stronger impact on behavior. Because systems often have several competing feedback loops operating simultaneously, those loops that dominate the system will determine the behavior.

At first, when fertility is higher than mortality, the reinforcing growth loop dominates the system and the resulting behavior is exponential growth. But that loop is gradually weakened as fertility falls. Finally, it exactly equals the strength of the balancing loop of mortality. At that point neither loop dominates, and we have dynamic equilibrium.

You saw shifting dominance in the thermostat system, when the outside temperature fell and the heat leaking out of the poorly insulated house overwhelmed the ability of the furnace to put heat into the room. Dominance shifted from the heating loop to the cooling loop.

> Complex behaviors of systems often arise as the relative strengths of feedback loops shift, causing first one loop and then another to dominate behavior.

There are only a few ways a population system could behave, and these depend on what happens to the "driving" variables, fertility and mortality. These are the only ones possible for a simple system of one reinforcing and one balancing loop. A stock governed by linked reinforcing and balancing loops will grow exponentially if the reinforcing loop dominates the balancing one. It will die off if the balancing loop dominates the reinforcing one. It will level off if the two loops are of equal strength (see Figure 25). Or it will do a sequence of these things, one after another, if the relative strengths of the two loops change over time (see Figure 26).

I chose some provocative population scenarios here to illustrate a point about models and the scenarios they can generate. Whenever you are confronted with a scenario (and you are, every time you hear about an economic prediction, a corporate budget, a weather forecast, future climate change, a stockbroker saying what is going to happen to a particular holding), there are questions you need to ask that will help you decide how good a representation of reality is the underlying model.

- Are the driving factors likely to unfold this way? (What are birth rate and death rate likely to do?)
- If they did, would the system react this way? (Do birth and death rates really cause the population stock to behave as we think it will?)
- What is driving the driving factors? (What affects birth rate? What affects death rate?)

The first question can't be answered factually. It's a guess about the future, and the future is inherently uncertain. Although you may have a strong opinion about it, there's no way to prove you're right until the future actually happens. A systems analysis can test a number of scenarios to see what happens if the driving factors do different things. That's usually one purpose

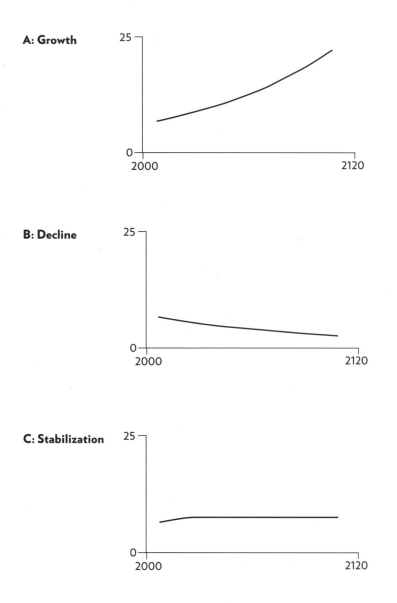

Figure 25. Three possible behaviors of a population: growth, decline, and stabilization.

of a systems analysis. But you have to be the judge of which scenario, if any, should be taken seriously as a future that might really be possible.

Dynamic systems studies usually are not designed to *predict* what will happen. Rather, they're designed to explore *what would happen,* if a number of driving factors unfold in a range of different ways.

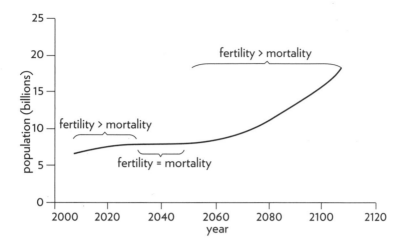

Figure 26. Shifting dominance of fertility and mortality loops.

The second question—whether the system really will react this way—is more scientific. It's a question about how good the model is. Does it capture the inherent dynamics of the system? Regardless of whether you think the driving factors *will* do that, *would the system behave like that* if they did?

In the population scenarios above, however likely you think they are, the answer to this second question is roughly yes, the population would behave like this, if the fertility and mortality did that. The population model I have used here is very simple. A more detailed model would distinguish age groups, for example. But basically this model responds the way a real population would, grow-

> System dynamics models explore possible futures and ask "what if" questions.

ing under the conditions when a real population would grow, declining when a real population would decline. The numbers are off, but the basic behavior pattern is realistic.

Finally, there is the third question. What is driving the driving factors?

QUESTIONS FOR TESTING THE VALUE OF A MODEL

1. Are the driving factors likely to unfold this way?
2. If they did, would the system react this way?
3. What is driving the driving factors?

> **Model utility depends not on whether its driving scenarios are realistic (since no one can know that for sure), but on whether it responds with a realistic pattern of behavior.**

What is adjusting the inflows and outflows? This is a question about system boundaries. It requires a hard look at those driving factors to see if they are actually independent, or if they are also embedded in the system.

Is there anything about the size of the population, for instance, that might feed back to influence fertility or mortality? Do other factors—economics, the environment, social trends—influence fertility and mortality? Does the size of the population affect those economic and environmental and social factors?

Of course, the answer to all of these questions is yes. Fertility and mortality are governed by feedback loops too. At least some of those feedback loops are themselves affected by the size of the population. This population "animal" is only one piece of a much larger system.[3]

One important piece of the larger system that affects population is the economy. At the heart of the economy is another reinforcing-loop-plus-balancing-loop system—the same kind of structure, with the same kinds of behavior, as the population (see Figure 27).

The greater the stock of physical capital (machines and factories) in the economy and the efficiency of production (output per unit of capital), the more output (goods and services) can be produced each year.

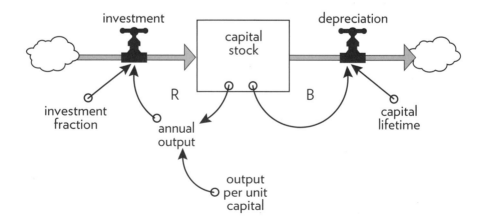

Figure 27. Like a living population, economic capital has a reinforcing loop (investment of output) governing growth and a balancing loop (depreciation) governing decline.

The more output that is produced, the more can be invested to make new capital. This is a reinforcing loop, like the birth loop for a population. The investment fraction is equivalent to the fertility. The greater the fraction of its output a society invests, the faster its capital stock will grow.

Physical capital is drained by depreciation—obsolescence and wearing-out. The balancing loop controlling depreciation is equivalent to the death loop in a population. The "mortality" of capital is determined by the average capital lifetime. The longer the lifetime, the smaller the fraction of capital that must be retired and replaced each year.

If this system has the same structure as the population system, it must have the same repertoire of behaviors. Over recent history world capital, like world population, has been dominated by its reinforcing loop and has been growing exponentially. Whether in the future it grows or stays constant or dies off depends on whether its reinforcing growth loop remains stronger than its balancing depreciation loop. That depends on

- the investment fraction—how much output the society invests rather than consumes,
- the efficiency of capital—how much capital it takes to produce a given amount of output, and
- the average capital lifetime.

If a constant fraction of output is reinvested in the capital stock and the efficiency of that capital (its ability to produce output) is also constant, the capital stock may decline, stay constant, or grow, depending on the lifetime of the capital. The lines in Figure 28 show systems with different average capital lifetimes. With a relatively short lifetime, the capital wears out faster than it is replaced. Reinvestment does not keep up with depreciation and the economy slowly declines. When depreciation just balances investment, the economy is in dynamic equilibrium. With a long lifetime, the capital stock grows exponentially. The longer the lifetime of capital, the faster it grows.

This is another example of a principle we've already encountered: You can make a stock grow by decreasing its outflow rate as well as by increasing its inflow rate!

Just as many factors influence the fertility and mortality of a population, so many factors influence the output ratio, investment fraction, and the lifetime of capital—interest rates, technology, tax policy, consumption

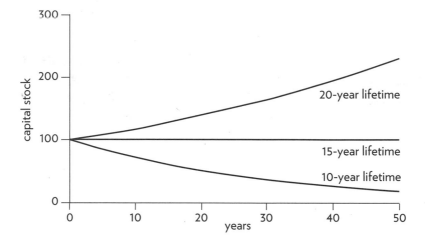

Figure 28. Growth in capital stock with changes in the lifetime of the capital. In a system with output per unit capital ratio of 1:3 and an investment fraction of 20 percent, capital with a lifetime of 15 years just keeps up with depreciation. A shorter lifetime leads to a declining stock of capital.

habits, and prices, to name just a few. Population itself influences investment, both by contributing labor to output, and by increasing demands on consumption, thereby decreasing the investment fraction. Economic output also feeds back to influence population in many ways. A richer economy usually has better health care and a lower death rate. A richer economy also usually has a lower birth rate.

In fact, just about any long-term model of a real economy should link together the two structures of population and capital to show how they affect each other. The central question of economic development is how to keep the reinforcing loop of capital accumulation from growing more slowly than the reinforcing loop of population growth—so that people are getting richer instead of poorer.[4]

It may seem strange to you that I call the capital system the same kind of "zoo animal" as the population system. A production system with factories and shipments and economic flows doesn't look much like a population system with babies being born and people aging and having more babies and dying. But from a systems point of view these systems, so dissimilar in many ways, have one important thing in common: their feedback-loop structures. Both have a stock governed by a reinforcing growth loop and a balancing death loop.

Systems with similar feedback structures produce similar dynamic behaviors.

Both also have an aging process. Steel mills and lathes and turbines get older and die just as people do.

One of the central insights of systems theory, as central as the observation that systems largely cause their own behavior, is that systems with similar feedback structures produce similar dynamic behaviors, even if the outward appearance of these systems is completely dissimilar.

A population is nothing like an industrial economy, except that both can reproduce themselves out of themselves and thus grow exponentially. And both age and die. A coffee cup cooling is like a warmed room cooling, and like a radioactive substance decaying, and like a population or industrial economy aging and dying. Each declines as the result of a balancing feedback loop.

A System with Delays—Business Inventory

Picture a stock of inventory in a store—a car dealership—with an inflow of deliveries from factories and an outflow of new car sales. By itself, this stock of cars on the dealership lot would behave like the water in a bathtub.

Now picture a regulatory feedback system designed to keep the inventory high enough so that it can always cover ten days' worth of sales (see Figure 29). The car dealer needs to keep some inventory because deliveries

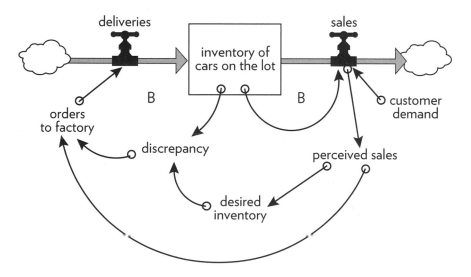

Figure 29. Inventory at a car dealership is kept steady by two competing balancing loops, one through sales and one through deliveries.

and purchases don't match perfectly every day. Customers make purchases that are unpredictable on a day-to-day basis. The car dealer also needs to provide herself with some extra inventory (a buffer) in case deliveries from suppliers are delayed occasionally.

The dealer monitors sales (perceived sales), and if, for example, they seem to be rising, she adjusts orders to the factory to bring inventory up to her new desired inventory that provides ten days' coverage at the higher sales rate. So, higher sales mean higher perceived sales, which means a higher discrepancy between inventory and desired inventory, which means higher orders, which will bring in more deliveries, which will raise inventory so it can comfortably supply the higher rate of sales.

This system is a version of the thermostat system—one balancing loop of sales draining the inventory stock and a competing balancing loop maintaining the inventory by resupplying what is lost in sales. Figure 30 shows the not very surprising result of an increase in consumer demand of 10 percent.

In Figure 31, I am putting something else into this simple model—three delays that are typical of what we experience in the real world.

First, there is a perception delay, intentional in this case. The car dealer doesn't react to just any blip in sales. Before she makes ordering decisions, she averages sales over the past five days to sort out real trends from temporary dips and spikes.

Second, there is a response delay. Even when it's clear that orders need

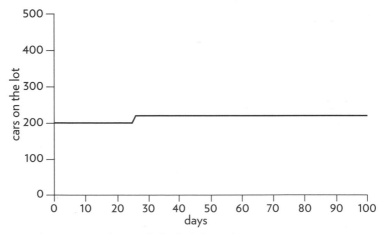

Figure 30. Inventory on the car dealership's lot with a permanent 10-percent increase in consumer demand starting on day 25.

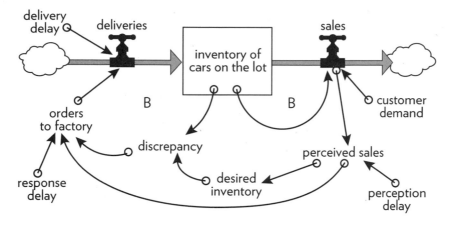

Figure 31. Inventory at a car dealership with three common delays now included in the picture—a perception delay, a response delay, and a delivery delay.

to be adjusted, she doesn't try to make up the whole adjustment in a single order. Rather, she makes up one-third of any shortfall with each order. Another way of saying that is, she makes partial adjustments over three days to be extra sure the trend is real. Third, there is a delivery delay. It takes five days for the supplier at the factory to receive an order, process it, and deliver it to the dealership.

Although this system still consists of just two balancing loops, like the simplified thermostat system, it doesn't behave like the thermostat system. Look at what happens, for example, as shown in Figure 32, when

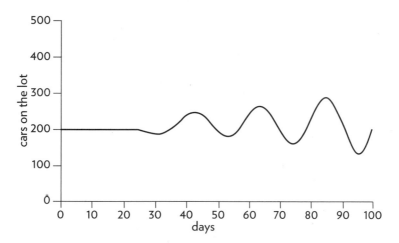

Figure 32. Response of inventory to a 10-percent increase in sales when there are delays in the system.

the business experiences the same permanent 10-percent jump in sales from an increase in customer demand.

Oscillations! A single step up in sales causes inventory to drop. The car dealer watches long enough to be sure the higher sales rate is going to last. Then she begins to order more cars to both cover the new rate of sales and bring the inventory up. But it takes time for the orders to come in. During that time inventory drops further, so orders have to go up a little more, to bring inventory back up to ten days' coverage.

Eventually, the larger volume of orders starts arriving, and inventory recovers—and more than recovers, because during the time of uncertainty about the actual trend, the owner has ordered too much. She now sees her mistake, and cuts back, but there are still high past orders coming in, so she orders even less. In fact, almost inevitably, since she still can't be sure of what is going to happen next, she orders too little. Inventory gets too low again. And so forth, through a series of oscillations around the new desired inventory level. As Figure 33 illustrates, what a difference a few delays make!

We'll see in a moment that there are ways to damp these oscillations in inventory, but first it's important to understand why they occur. It isn't because the car dealer is stupid. It's because she is struggling to operate in a system in which she doesn't have, and can't have, timely information and in which physical delays prevent her actions from having an imme-diate effect on inventory. She doesn't know what her customers will do

A delay in a balancing feedback loop makes a system likely to oscillate.

next. When they do something, she's not sure they'll keep doing it. When she issues an order, she doesn't see an immediate response. This situation of information insufficiency and physical delays is very common. Oscillations like these are frequently encountered in inventories and in many other systems. Try taking a shower sometime where there's a very long pipe between the hot- and cold-water mixer and the showerhead, and you'll experience directly the joys of hot and cold oscillations because of a long response delay.

How much of a delay causes what kind of oscillation under what circum-stances is not a simple matter. I can use this inventory system to show you why.

"These oscillations are intolerable," says the car dealer (who is herself a learning system, determined now to change the behavior of the inventory

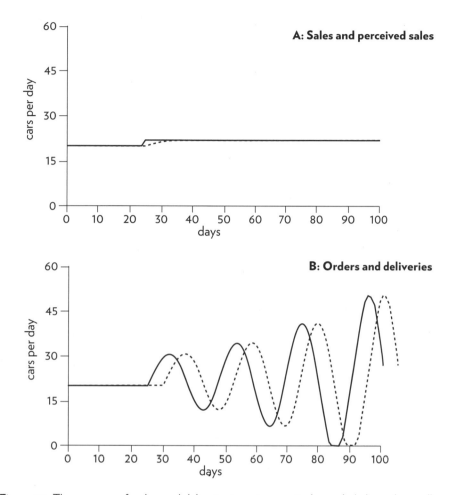

Figure 33. The response of orders and deliveries to an increase in demand. *A* shows the small but sharp step up in sales on day 25 and the car dealer's "perceived" sales, in which she averages the change over 3 days. *B* shows the resulting ordering pattern, tracked by the actual deliveries from the factory.

system). "I'm going to shorten the delays. There's not much I can do about the delivery delay from the factory, so I'm going to react faster myself. I'll average sales trends over only two days instead of five before I make order adjustments."

Figure 34 illustrates what happens when the dealer's perception delay is shortened from five days to two.

Not much happens when the car dealer shortens her perception delay. If anything the oscillations in the inventory of cars on the lot are a bit worse. And if, instead of shortening her perception time, the car dealer tries shortening her reaction time—making up perceived shortfalls in two days instead of three—things get very much worse, as shown in Figure 35.

Something has to change and, since this system has a learning person within it, something will change. "High leverage, wrong direction," the system-thinking car dealer says to herself as she watches this failure of a policy intended to stabilize the oscillations. This perverse kind of result can

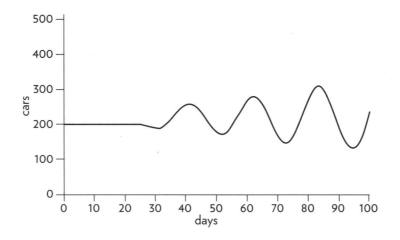

Figure 34. The response of inventory to the same increase in demand with a shortened perception delay.

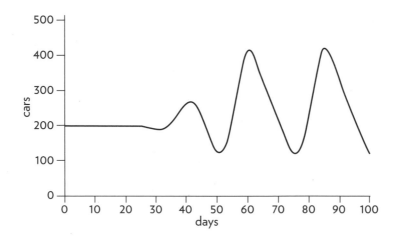

Figure 35. The response of inventory to the same increase in demand with a shortened reaction time. Acting faster makes the oscillations worse!

be seen all the time—someone trying to fix a system is attracted intuitively
to a policy lever that in fact does have a strong effect on the system. And
then the well-intentioned fixer pulls the lever in the wrong direction! This
is just one example of how we can be surprised by the counterintuitive
behavior of systems when we start trying to change them.

Part of the problem here is that the car dealer has been reacting not too
slowly, but too quickly. Given the configuration of this system, she has been
overreacting. Things would go better if, instead of decreasing her response
delay from three days to two, she would increase the delay from three days
to six, as illustrated in Figure 36.

As Figure 36 shows, the oscillations are greatly damped with this change,
and the system finds its new equilibrium fairly
efficiently.

> Delays are pervasive in systems, and they are strong determinants of behavior. **Changing the length of a delay may** (or may not, depending on the type of delay and the relative lengths of other delays) **make a large change in the behavior of a system.**

The most important delay in this system is
the one that is not under the direct control of
the car dealer. It's the delay in delivery from the
factory. But even without the ability to change
that part of her system, the dealer can learn to
manage inventory quite well.

Changing the delays in a system can make it
much easier or much harder to manage. You
can see why system thinkers are somewhat fanatic on the subject of delays.
We're always on the alert to see where delays occur in systems, how long

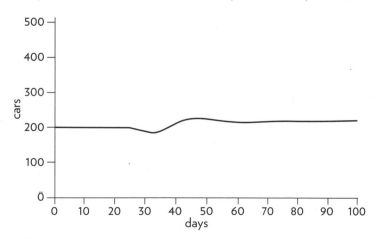

Figure 36. The response of inventory to the same increase in demand with a slowed reaction
time.

they are, whether they are delays in information streams or in physical processes. We can't begin to understand the dynamic behavior of systems unless we know where and how long the delays are. And we are aware that some delays can be powerful policy levers. Lengthening or shortening them can produce major changes in the behavior of systems.

In the big picture, one store's inventory problem may seem trivial and fixable. But imagine that the inventory is that of all the unsold automobiles in America. Orders for more or fewer cars affect production not only at assembly plants and parts factories, but also at steel mills, rubber and glass plants, textile producers, and energy producers. Everywhere in this system are perception delays, production delays, delivery delays, and construction delays. Now consider the link between car production and jobs—increased production increases the number of jobs allowing more people to buy cars. That's a reinforcing loop, which also works in the opposite direction— less production, fewer jobs, fewer car sales, less production. Put in another reinforcing loop, as speculators buy and sell shares in the auto and auto-supply companies based on their recent performance, so that an upsurge in production produces an upsurge in stock price, and vice versa.

That very large system, with interconnected industries responding to each other through delays, entraining each other in their oscillations, and being amplified by multipliers and speculators, is the primary cause of business cycles. Those cycles don't come from presidents, although presidents can do much to ease or intensify the optimism of the upturns and the pain of the downturns. Economies are extremely complex systems; they are full of balancing feedback loops with delays, and they are inherently oscillatory.[5]

Two-Stock Systems

A Renewable Stock Constrained by a Nonrenewable Stock—an Oil Economy

The systems I've displayed so far have been free of constraints imposed by their surroundings. The capital stock of the industrial economy model didn't require raw materials to produce output. The population didn't need food. The thermostat-furnace system never ran out of oil. These simple models of the systems have been able to operate according to their uncon-strained internal dynamics, so we could see what those dynamics are.

But any real physical entity is always surrounded by and exchanging things with its environment. A corporation needs a constant supply of energy and materials and workers and managers and customers. A growing corn crop needs water and nutrients and protection from pests. A population needs food and water and living space, and if it's a human population, it needs jobs and education and health care and a multitude of other things. Any entity that is using energy and processing materials needs a place to put its wastes, or a process to carry its wastes away.

Therefore, any physical, growing system is going to run into some kind of constraint, sooner or later. That constraint will take the form of a balancing loop that in some way shifts the dominance of the reinforcing loop driving the growth behavior, either by strengthening the outflow or by weakening the inflow.

Growth in a constrained environment is very common, so common that systems thinkers call it the "limits-to-growth" archetype. (We'll explore more archetypes—frequently found system structures that produce familiar behavior patterns—in Chapter Five.) Whenever we see a growing entity, whether it be a population, a corporation, a bank account, a rumor, an epidemic, or sales of a new product, we look for the reinforcing loops that are driving it and for the balancing loops that ultimately will constrain it. We know those balancing loops are there, even if they are not yet dominating the system's behavior, because no real physical system can grow forever. Even a hot new product will saturate the market eventually. A chain reaction in a nuclear power plant or bomb will run out of fuel. A virus will run out of susceptible people to infect. An economy may be constrained by physical capital or monetary capital or labor or markets or management or resources or pollution.

> In physical, exponentially growing systems, there must be at least one reinforcing loop driving the growth *and* at least one balancing loop constraining the growth, because no physical system can grow forever in a finite environment.

Like resources that supply the inflows to a stock, a pollution constraint can be renewable or nonrenewable. It's nonrenewable if the environment has no capacity to absorb the pollutant or make it harmless. It's renewable if the environment has a finite, usually variable, capacity for removal. Everything said here about resource-constrained systems, therefore, applies with the same dynamics but opposite flow directions to pollution-constrained systems.

The limits on a growing system may be temporary or permanent. The system may find ways to get around them for a short while or a long while, but eventually there must come some kind of accommodation, the system adjusting to the constraint, or the constraint to the system, or both to each other. In that accommodation come some interesting dynamics.

Whether the constraining balancing loops originate from a renewable or nonrenewable resource makes some difference, not in whether growth can continue forever, but in how growth is likely to end.

Let's look, to start, at a capital system that makes its money by extracting a nonrenewable resource—say an oil company that has just discovered a huge new oil field. See Figure 37.

The diagram in Figure 37 may look complicated, but it's no more than a capital-growth system like the one we've already seen, using "profit" instead of "output." Driving depreciation is the now-familiar balancing loop: the more capital stock, the more machines and refineries there are

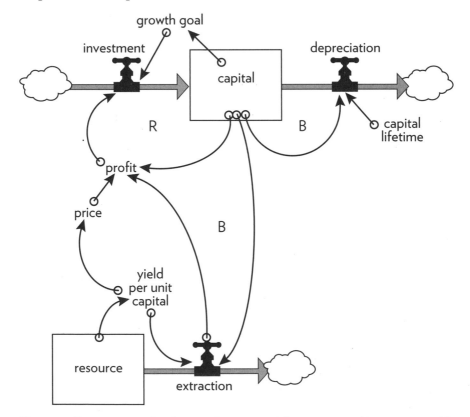

Figure 37. Economic capital, with its reinforcing growth loop constrained by a nonrenewable resource.

that fall apart and wear out, reducing the stock of capital. In this example, the capital stock of oil drilling and refining equipment depreciates with a 20-year lifetime—meaning 1/20 (or 5 percent) of the stock is taken out of commission each year. It builds itself up through investment of profits from oil extraction. So we see the reinforcing loop: More capital allows more resource extraction, creating more profits that can be reinvested. I've assumed that the company has a goal of 5 percent annual growth in its business capital. If there isn't enough profit for 5 percent growth, the company invests whatever profits it can.

Profit is income minus cost. Income in this simple representation is just the price of oil times the amount of oil the company extracts. Cost is equal to capital times the operating cost (energy, labor, materials, etc.) per unit of capital. For the moment, I'll make the simplifying assumptions that both price and operating cost per unit of capital are constant.

What is not assumed to be constant is the yield of resource per unit of capital. Because this resource is not renewable, as in the case of oil, the stock feeding the extraction flow does not have an input. As the resource is extracted—as an oil well is depleted—the next barrel of oil becomes harder to get. The remaining resource is deeper down, or more dilute, or in the case of oil, under less natural pressure to force it to the surface. More and more costly and technically sophisticated measures are required to keep the resource coming.

Here is a new balancing feedback loop that ultimately will control the growth of capital: the more capital, the higher the extraction rate. The higher the extraction rate, the lower the resource stock. The lower the resource stock, the lower the yield of resource per unit of capital, so the lower the profit (with price assumed constant) and the lower the invest- ment rate—therefore, the lower the rate of growth of capital. I could assume that resource depletion feeds back through operating cost as well as capital efficiency. In the real world it does both. In either case, the ensuing behavior pattern is the same—the classic dynamics of depletion (see Figure 38).

The system starts out with enough oil in the underground deposit to supply the initial scale of operation for 200 years. But, actual extraction peaks at about 40 years because of the surprising effect of exponential growth in extraction. At an investment rate of 10 percent per year, the capi- tal stock and therefore the extraction rate both grow at 5 percent per year and so double in the first 14 years. After 28 years, while the capital stock has

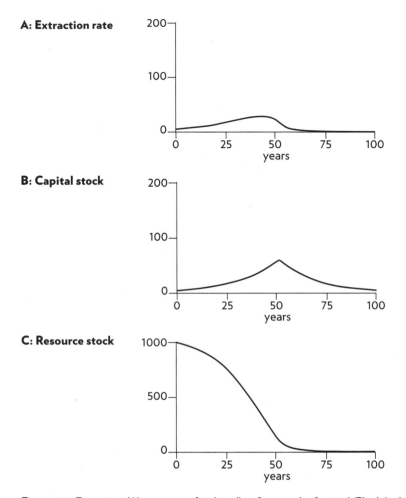

Figure 38. Extraction (A) creates profits that allow for growth of capital (B) while depleting the nonrenewable resource (C). The greater the accumulation of capital, the faster the resource is depleted.

quadrupled, extraction is starting to lag because of falling yield per unit of capital. By year 50 the cost of maintaining the capital stock has overwhelmed the income from resource extraction, so profits are no longer sufficient to keep investment ahead of depreciation. The operation quickly shuts down, as the capital stock declines. The last and most expensive of the resource stays in the ground; it doesn't pay to get it out.

What happens if the original resource turns out to be twice as large as the geologists first thought—or four times as large? Of course, that makes a huge difference in the total amount of oil that can be extracted from this field. But with the continued goal of 10 percent per year reinvestment producing 5 percent per year capital growth, each doubling of the resource makes a difference of only about 14 years in the timing of the peak extraction rate, and in the lifetime of any jobs or communities dependent on the extraction industry (see Figure 39).

A quantity growing exponentially toward a constraint or limit reaches that limit in a surprisingly short time.

The higher and faster you grow, the farther and faster you fall, when you're building up a capital stock dependent on a nonrenewable resource. In the face of exponential growth of extraction or use, a doubling or quadrupling of the nonrenewable resource give little added time to develop alternatives.

If your concern is to extract the resource and make money at the maximum possible rate, then the ultimate size of the resource is the most important number in this system. If, say, you're a worker at the mine or oil field, and your concern is with the lifetime of your job and stability of your community, then there are two important numbers: the size of the resource and the desired growth rate of capital. (Here is a good example of the goal of a feedback loop being crucial to the behavior of a system.) The real choice in the management of a nonrenewable resource is whether to get rich very fast or to get less rich but stay that way longer.

The graph in Figure 40 shows the development of the extraction rate over time, given desired growth rates above depreciation varying from 1 percent annually, to 3 percent, 5 percent, and 7 percent. With a 7 percent growth rate, extraction of this "200-year supply" peaks within 40 years.

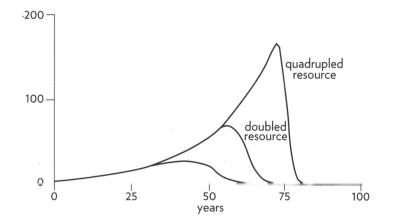

Figure 39. Extraction with two times or four times as large a resource to draw on. Each doubling of the resource makes a difference of only about fourteen years in the peak of extraction.

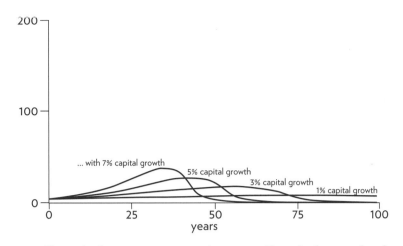

Figure 40. The peak of extraction comes much more quickly as the fraction of profits reinvested increases.

Imagine the effects of this choice not only on the profits of the company, but on the social and natural environments of the region.

Earlier I said I would make the simplifying assumption that price was constant. But what if that's not true? Suppose that in the short term the resource is so vital to consumers that a higher price won't decrease demand. In that case, as the resource gets scarce and price rises steeply, as shown in Figure 41.

The higher price gives the industry higher profits, so investment goes up, capital stock continues rising, and the more costly remaining resources can be extracted. If you compare Figure 41 with Figure 38, where price was held constant, you can see that the main effect of rising price is to build the capital stock higher before it collapses.

The same behavior results, by the way, if prices don't go up but if technology brings operating costs down—as has actually happened, for example, with advanced recovery techniques from oil wells, with the beneficiation process to extract low-grade taconite from exhausted iron mines, and with the cyanide leaching process that allows profitable extraction even from the tailings of gold and silver mines.

We all know that individual mines and fossil fuel deposits and groundwater aquifers can be depleted. There are abandoned mining towns and oil fields all over the world to testify to the reality of the behavior we've seen here. Resource companies understand this dynamic too. Well before depletion makes capital less efficient in one place, companies shift investment

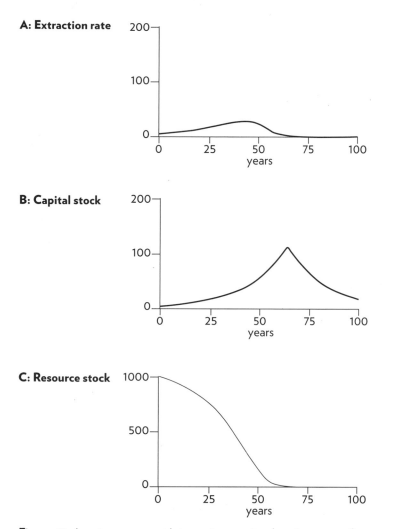

Figure 41. As price goes up with increasing scarcity, there is more profit to reinvest, and the capital stock can grow larger (B) driving extraction up for longer (A). The consequence is that the resource (C) is depleted even faster at the end.

to discovery and development of another deposit somewhere else. But, if there are local limits, eventually will there be global ones?

I'll leave you to have this argument with yourself, or with someone of the opposite persuasion. I will just point out that, according to the dynamics of depletion, the larger the stock of initial resources, the more new discoveries, the longer the growth loops elude the control loops, and the higher the capital stock and its extraction rate grow, and the earlier, faster, and farther will be the economic fall on the back side of the production peak.

Unless, perhaps, the economy can learn to operate entirely from renewable resources.

Renewable Stock Constrained by a Renewable Stock—a Fishing Economy

Assume the same capital system as before, except that now there is an inflow to the resource stock, making it renewable. The renewable resource in this system could be fish and the capital stock could be fishing boats. It also could be trees and sawmills, or pasture and cows. Living renewable resources such as fish or trees or grass can regenerate themselves from themselves with a reinforcing feedback loop. Nonliving renewable resources such as sunlight or wind or water in a river are regenerated not through a reinforcing loop, but through a steady input that keeps refilling the resource stock no matter what the current state of that stock might be. This same "renewable resource system" structure occurs in an epidemic of a cold virus. It spares its victims who are then able to catch another cold. Sales of a product people need to buy regularly is also a renewable resource system; the stock of potential customers is ever regenerated. Likewise an insect infestation that destroys part but not all of a plant; the plant can regenerate and the insect can eat more. In all these cases, there is an input that keeps refilling the constraining resource stock (as shown in Figure 42).

We will use the example of a fishery. Once again, assume that the lifetime of capital is 20 years and the industry will grow, if it can, at 5 percent per year. As with the nonrenewable resource, assume that as the resource gets scarce it costs more, in terms of capital, to harvest it. Bigger fishing boats that can go longer distances and are equipped with sonar are needed to find the last schools of fish. Or miles-long drift nets are needed to catch them. Or on-board refrigeration systems are needed to bring them back to port from longer distances. All this takes more capital.

The regeneration rate of the fish is not constant, but is dependent on the number of fish in the area—fish density. If the fish are very dense, their reproduction rate is near zero, limited by available food and habitat. If the fish population falls a bit, it can regenerate at a faster and faster rate, because it can take advantage of unused nutrients or space in the ecosystem. But at some point the fish reproduction rate reaches its maximum. If the population is further depleted, it breeds not faster and faster, but slower and slower. That's because the fish can't find each other, or because another species has moved into its niche.

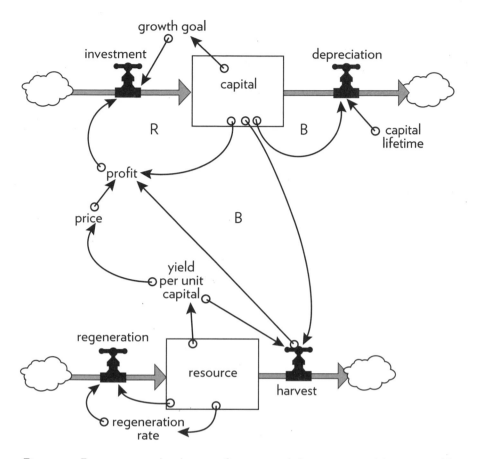

Figure 42. Economic capital with its reinforcing growth loop constrained by a renewable resource.

This simplified model of a fishery economy is affected by three nonlinear relationships: price (scarcer fish are more expensive); regeneration rate (scarcer fish don't breed much, nor do crowded fish); and yield per unit of capital (efficiency of the fishing technology and practices).

This system can produce many different sets of behaviors. Figure 43 shows one of them.

In Figure 43, we see capital and fish harvest rise exponentially at first. The fish population (the resource stock) falls, but that stimulates the fish reproduction rate. For decades the resource can go on supplying an exponentially increasing harvest rate. Eventually, the harvest rises too far and the fish population falls low enough to reduce the profitability of the fishing fleet. The balancing feedback of falling harvest reducing profits brings down the investment rate quickly enough to bring the fishing fleet into equilibrium with the fish resource. The fleet can't grow forever, but it can maintain a high and steady harvest rate forever.

Just a minor change in the strength of the controlling balancing feedback loop through yield per unit of capital, however, can make a surprising difference. Suppose that in an attempt to raise the catch in the fishery,

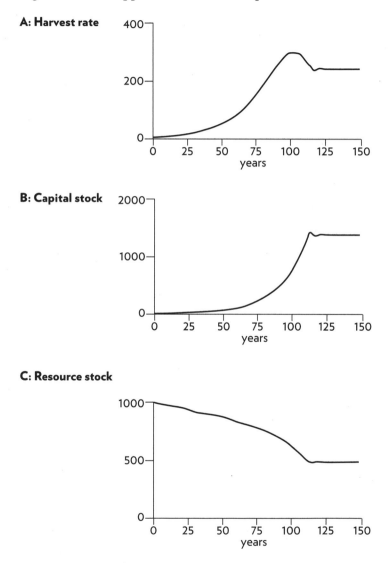

Figure 43. Annual harvest (A) creates profits that allow for growth of capital stock (B), but the harvest levels off, after a small overshoot in this case. The result of leveling harvest is that the resource stock (C) also stabilizes.

the industry comes up with a technology to improve the efficiency of the boats (sonar, for example, to find the scarcer fish). As the fish population declines, the fleet's ability to pull in the same catch per boat is maintained just a little longer (see Figure 44).

Figure 44 shows another case of high leverage, wrong direction! This

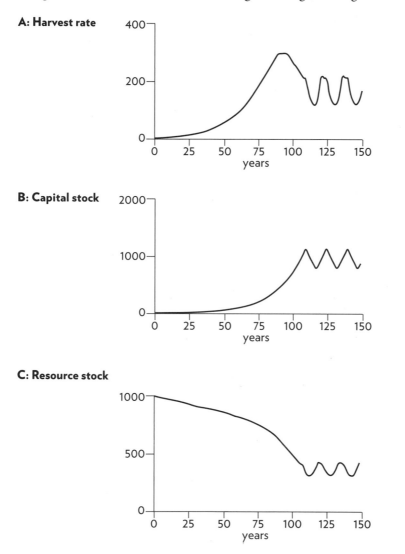

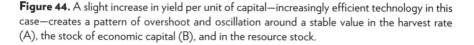

Figure 44. A slight increase in yield per unit of capital—increasingly efficient technology in this case—creates a pattern of overshoot and oscillation around a stable value in the harvest rate (A), the stock of economic capital (B), and in the resource stock.

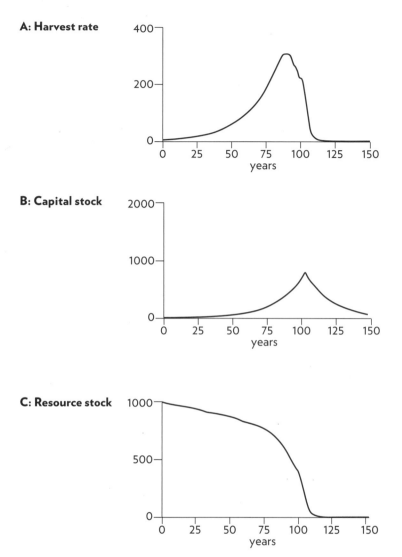

Figure 45. An even greater increase in yield per unit of capital creates a patterns of overshoot and collapse in the harvest (A), the economic capital (B), and the resource (C).

technical change, which increases the productivity of all fishermen, throws the system into instability. Oscillations appear!

If the fishing technology gets even better, the boats can go on operating economically even at very low fish densities. The result can be a nearly complete wipeout both of the fish and of the fishing industry. The consequence is the marine equivalent of desertification. The fish have been turned, for all practical purposes, into a nonrenewable resource. Figure 45 illustrates this scenario.

In many real economies based on real renewable resources—as opposed to this simple model—the very small surviving population retains the potential to build its numbers back up again, once the capital driving the harvest is gone. The whole pattern is repeated, decades later. Very long-term renewable-resource cycles like these have been observed, for example, in the logging industry in New England, now in its third cycle of growth, overcutting, collapse, and eventual regeneration of the resource. But this is not true for all resource populations. More and more, increases in technology and harvest efficiency have the ability to drive resource populations to extinction.

Nonrenewable resources are *stock*-**limited.** The entire stock is available at once, and can be extracted at any rate (limited mainly by extraction capital). But since the stock is not renewed, the faster the extraction rate, the shorter the lifetime of the resource.

Renewable resources are *flow*-**limited.** They can support extraction or harvest indefinitely, but only at a finite flow rate equal to their regeneration rate. If they are extracted faster than they regenerate, they may eventually be driven below a critical threshold and become, for all practical purposes, nonrenewable.

Whether a real renewable resource system can survive overharvest depends on what happens to it during the time when the resource is severely depleted. A very small fish population may become especially vulnerable to pollution or storms or lack of genetic diversity. If this is a forest or grassland resource, the exposed soils may be vulnerable to erosion. Or the nearly empty ecological niche may be filled in by a competitor. Or perhaps the depleted resource can survive and rebuild itself again.

I've shown three sets of possible behaviors of this renewable resource system here:

- overshoot and adjustment to a sustainable equilibrium,
- overshoot beyond that equilibrium followed by oscillation around it, and
- overshoot followed by collapse of the resource and the industry dependent on the resource.

Which outcome actually occurs depends on two things. The first is the critical threshold beyond which the resource population's ability to regenerate itself is damaged. The second is the rapidity and effectiveness of the balancing feedback loop that slows capital growth as the resource becomes depleted. If the feedback is fast enough to stop capital growth before the critical threshold is reached, the whole system comes smoothly into equilibrium. If the balancing feedback is slower and less effective, the system oscillates. If the balancing loop is very weak, so that capital can go on growing even as the resource is reduced below its threshold ability to regenerate itself, the resource and the industry both collapse.

Neither renewable nor nonrenewable limits to growth allow a physical stock to grow forever, but the constraints they impose are dynamically quite different. The difference comes because of the difference between stocks and flows.

The trick, as with all the behavioral possibilities of complex systems, is to recognize what structures contain which latent behaviors, and what conditions release those behaviors—and, where possible, to arrange the structures and conditions to reduce the probability of destructive behaviors and to encourage the possibility of beneficial ones.

PART TWO
Systems and Us

THREE

Why Systems Work So Well

If the land mechanism as a whole is good, then every part is good, whether we understand it or not. If the biota, in the course of aeons, has built something we like but do not understand, then who but a fool would discard seemingly useless parts? To keep every cog and wheel is the first precaution of intelligent tinkering.

—Aldo Leopold,[1] forester

Chapter Two introduced simple systems that create their own behavior based on their structures. Some are quite elegant—surviving the buffeting of the world—and, within limits, regaining their composure and proceeding on about their business of maintaining a room's temperature, depleting an oil field, or bringing into balance the size of a fishing fleet with the productivity of a fishery resource.

If pushed too far, systems may well fall apart or exhibit heretofore unobserved behavior. But, by and large, they manage quite well. And that is the beauty of systems: They can work so well. When systems work well, we see a kind of harmony in their functioning. Think of a community kicking in to high gear to respond to a storm. People work long hours to help victims, talents and skills emerge; once the emergency is over, life goes back to "normal."

Why do systems work so well? Consider the properties of highly functional systems—machines or human communities or ecosystems—which are familiar to you. Chances are good that you may have observed one of three characteristics: resilience, self-organization, or hierarchy.

Resilience

Placing a system in a straitjacket of constancy can cause fragility to evolve.

—C. S. Holling,[2] ecologist

Resilience has many definitions, depending on the branch of engineering, ecology, or system science doing the defining. For our purposes, the normal dictionary meaning will do: "the ability to bounce or spring back into shape, position, etc., after being pressed or stretched. Elasticity. The ability to recover strength, spirits, good humor, or any other aspect quickly." Resilience is a measure of a system's ability to survive and persist within a variable environment. The opposite of resilience is brittleness or rigidity.

Resilience arises from a rich structure of many feedback loops that can work in different ways to restore a system even after a large perturbation. A single balancing loop brings a system stock back to its desired state. Resilience is provided by several such loops, operating through different mechanisms, at different time scales, and with redundancy—one kicking in if another one fails.

A set of feedback loops that can *restore or rebuild feedback loops* is resilience at a still higher level—meta-resilience, if you will. Even higher meta-meta-resilience comes from feedback loops that can *learn, create, design, and evolve* ever more complex restorative structures. Systems that can do this are self-organizing, which will be the next surprising system characteristic I come to.

The human body is an astonishing example of a resilient system. It can fend off thousands of different kinds of invaders, it can tolerate wide ranges of temperature and wide variations in food supply, it can reallocate blood supply, repair rips, gear up or slow down metabolism, and compensate to some extent for missing or defective parts. Add to it a self-organizing intelligence that can learn, socialize, design technologies, and even transplant body parts, and you have a formidably resilient system—although not infinitely so, because, so far at least, no human body-plus-intelligence has been resilient enough to keep itself or any other body from eventually dying.

> There are always limits to resilience.

Ecosystems are also remarkably resilient, with multiple species holding each other in check, moving around in space, multiplying or declining over time in response to weather and the availability of nutrients and the impacts of human activities. Populations and ecosystems also have the ability to "learn" and evolve through their incredibly rich genetic variability. They can, given enough time, come up with whole new systems to take advantage of changing opportunities for life support.

Resilience is not the same thing as being static or constant over time. Resilient systems can be very dynamic. Short-term oscillations, or periodic outbreaks, or long cycles of succession, climax, and collapse may in fact be the normal condition, which resilience acts to restore!

And, conversely, systems that are constant over time can be unresilient. This distinction between static stability and resilience is important. Static stability is something you can see; it's measured by variation in the condition of a system week by week or year by year. Resilience is something that may be very hard to see, unless you exceed its limits, overwhelm and damage the balancing loops, and the system structure breaks down. Because resilience may not be obvious without a whole-system view, people often sacrifice resilience for stability, or for productivity, or for some other more immediately recognizable system property.

- Injections of genetically engineered bovine growth hormone increase the milk production of a cow without proportionately increasing the cow's food intake. The hormone diverts some of the cow's metabolic energy from other bodily functions to milk production. (Cattle breeding over centuries has done much the same thing but not to the same degree.) The cost of increased production is lowered resilience. The cow is less healthy, less long-lived, more dependent on human management.
- Just-in-time deliveries of products to retailers or parts to manufacturers have reduced inventory instabilities and brought down costs in many industries. The just-in-time model also has made the production system more vulnerable, however, to perturbations in fuel supply, traffic flow, computer breakdown, labor availability, and other possible glitches.
- Hundreds of years of intensive management of the forests of Europe gradually have replaced native ecosystems with single-age, single-species plantations, often of nonnative trees. These

forests are designed to yield wood and pulp at a high rate
indefinitely. However, without multiple species interacting
with each other and drawing and returning varying combinations of nutrients from the soil, these forests have lost their
resilience. They seem to be especially vulnerable to a new form
of insult: industrial air pollution.

Many chronic diseases, such as cancer and heart disease, come from
breakdown of resilience mechanisms that repair DNA, keep blood vessels
flexible, or control cell division. Ecological disasters in many places come
from loss of resilience, as species are removed from ecosystems, soil chemistry and biology are disturbed, or toxins build up. Large organizations of
all kinds, from corporations to governments, lose their resilience simply
because the feedback mechanisms by which they sense and respond to
their environment have to travel through too many layers of delay and
distortion. (More on that in a minute, when we come to hierarchies.)

I think of resilience as a plateau upon which the system can play, perform-
ing its normal functions in safety. A resilient system has a big plateau, a
lot of space over which it can wander, with gentle, elastic walls that will
bounce it back, if it comes near a dangerous edge.
As a system loses its resilience, its plateau shrinks,
and its protective walls become lower and more
rigid, until the system is operating on a knife-
edge, likely to fall off in one direction or another
whenever it makes a move. Loss of resilience can
come as a surprise, because the system usually is
paying much more attention to its play than to its
playing space. One day it does something it has
done a hundred times before and crashes.

> **Systems need to be managed not only for productivity or stability, they also need to be managed for resilience—** the ability to recover from perturbation, the ability to restore or repair themselves.

Awareness of resilience enables one to see many ways to preserve or
enhance a system's own restorative powers. That awareness is behind the
encouragement of natural ecosystems on farms, so that predators can
take on more of the job of controlling pests. It is behind "holistic" health
care that tries not only to cure disease but also to build up a body's internal resistance. It is behind aid programs that do more than give food or
money—that try to change the circumstances that obstruct peoples' ability
to provide their own food or money.

Self-Organization

[Evolution] appears to be not a series of accidents the course of which is determined only by the change of environments during earth history and the resulting struggle for existence, . . . but is governed by definite laws. . . . The discovery of these laws constitutes one of the most important tasks of the future.

—Ludwig von Bertalanffy,[3] biologist

The most marvelous characteristic of some complex systems is their ability to learn, diversify, complexify, evolve. It is the ability of a single fertilized ovum to generate, out of itself, the incredible complexity of a mature frog, or chicken, or person. It is the ability of nature to have diversified millions of fantastic species out of a puddle of organic chemicals. It is the ability of a society to take the ideas of burning coal, making steam, pumping water, and specializing labor, and develop them eventually into an automobile assembly plant, a city of skyscrapers, a worldwide network of communications.

This capacity of a system to make its own structure more complex is called **self-organization**. You see self-organization in a small, mechanistic way whenever you see a snowflake, or ice feathers on a poorly insulated window, or a supersaturated solution suddenly forming a garden of crystals. You see self-organization in a more profound way whenever a seed sprouts, or a baby learns to speak, or a neighborhood decides to come together to oppose a toxic waste dump.

Self-organization is such a common property, particularly of living systems, that we take it for granted. If we didn't, we would be dazzled by the unfolding systems of our world. And if we weren't nearly blind to the property of self-organization, we would do better at encouraging, rather than destroying, the self-organizing capacities of the systems of which we are a part.

Like resilience, self-organization is often sacrificed for purposes of short-term productivity and stability. Productivity and stability are the usual excuses for turning creative human beings into mechanical adjuncts to production processes. Or for narrowing the genetic variability of crop plants. Or for establishing bureaucracies and theories of knowledge that treat people as if they were only numbers.

Self-organization produces heterogeneity and unpredictability. It is likely to come up with whole new structures, whole new ways of doing things. It requires freedom and experimentation, and a certain amount of disorder. These conditions that encourage self-organization often can be scary for individuals and threatening to power structures. As a consequence, education systems may restrict the creative powers of children instead of stimulating those powers. Economic policies may lean toward supporting established, powerful enterprises rather than upstart, new ones. And many governments prefer their people not to be too self-organizing.

Fortunately, self-organization is such a basic property of living systems that even the most overbearing power structure can never fully kill it, although in the name of law and order, self-organization can be suppressed for long, barren, cruel, boring periods.

Systems theorists used to think that self-organization was such a complex property of systems that it could never be understood. Computers were used to model mechanistic, "deterministic" systems, not evolutionary ones, because it was suspected, without much thought, that evolutionary systems were simply not understandable.

New discoveries, however, suggest that just a few simple organizing principles can lead to wildly diverse self-organizing structures. Imagine a triangle with three equal sides. Add to the middle of each side another equilateral triangle, one-third the size of the first one. Add to each of the new sides another triangle, one-third smaller. And so on. The result is called a Koch snowflake. (See Figure 46.) Its edge has tremendous length—but it can be contained within a circle. This structure is one simple example of fractal geometry—a realm of mathematics and art populated by elaborate shapes formed by relatively simple rules.

Similarly, the delicate, beautiful, intricate structure of a stylized fern can be generated by a computer with just a few simple fractal rules. The

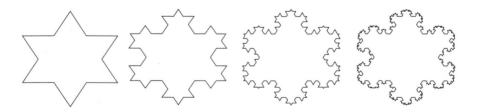

Figure 46. Even a delicate and intricate pattern, such as the Koch snowflake shown here, can evolve from a simple set of organizing principles or decision rules.

differentiation of a single cell into a human being probably proceeds by some similar set of geometric rules, basically simple, but generating utter complexity. (It is because of fractal geometry that the average human lung has enough surface area to cover a tennis court.)

Here are some other examples of simple organizing rules that have led to self-organizing systems of great complexity:

- All of life, from viruses to redwood trees, from amoebas to elephants, is based on the basic organizing rules encapsulated in the chemistry of DNA, RNA, and protein molecules.
- The agricultural revolution and all that followed started with the simple, shocking ideas that people could stay settled in one place, own land, select and cultivate crops.
- "God created the universe with the earth at its center, the land with the castle at its center, and humanity with the Church at its center"—the organizing principle for the elaborate social and physical structures of Europe in the Middle Ages.
- "God and morality are outmoded ideas; people should be objective and scientific, should own and multiply the means of production, and should treat people and nature as instrumental inputs to production"—the organizing principles of the Industrial Revolution.

Out of simple rules of self-organization can grow enormous, diversifying crystals of technology, physical structures, organizations, and cultures.

Science knows now that self-organizing systems can arise from simple rules. Science, itself a self-organizing system, likes to think that all the complexity of the world must arise, ultimately, from simple rules. Whether that actually happens is something that science does not yet know.

> **Systems often have the property of self-organization—the ability to structure themselves, to create new structure, to learn, diversify, and complexify.** Even complex forms of self-organization may arise from relatively simple organizing rules—or may not.

Hierarchy

So, naturalists observe, a flea
Has smaller Fleas that on him prey;
And these have smaller still to bite 'em,
And so proceed *ad infinitum.*

—Jonathan Swift,[4] 18th century poet

In the process of creating new structures and increasing complexity, one thing that a self-organizing system often generates is **hierarchy.**

The world, or at least the parts of it humans think they understand, is organized in subsystems aggregated into larger subsystems, aggregated into still larger subsystems. A cell in your liver is a subsystem of an organ, which is a subsystem of you as an organism, and you are a subsystem of a family, an athletic team, a musical group, and so forth. These groups are subsystems of a town or city, and then a nation, and then the whole global socioeconomic system that dwells within the biosphere system. This arrangement of systems and subsystems is called a hierarchy.

Corporate systems, military systems, ecological systems, economic systems, living organisms, are arranged in hierarchies. It is no accident that that is so. If subsystems can largely take care of themselves, regulate themselves, maintain themselves, and yet serve the needs of the larger system, while the larger system coordinates and enhances the functioning of the subsystems, a stable, resilient, and efficient structure results. It is hard to imagine how any other kind of arrangement could have come to be.

INTERLUDE • *Why the Universe Is Organized into Hierarchies—a Fable*

There once were two watchmakers, named Hora and Tempus. Both of them made fine watches, and they both had many customers. People dropped into their stores, and their phones rang constantly with new orders. Over the years, however, Hora prospered, while Tempus became poorer and poorer. That's because Hora discovered the principle of hierarchy. . . .

The watches made by both Hora and Tempus consisted of about one thousand parts each. Tempus put his together in such a way that if he had one partly assembled and had to put it down—to answer the phone, say—it fell to pieces. When he came back to it, Tempus would have to start all over again. The more his customers phoned him, the harder it became for him to find enough uninterrupted time to finish a watch.

Hora's watches were no less complex than those of Tempus, but he put together stable subassemblies of about ten elements each. Then he put ten of these subassemblies together into a larger assembly; and ten of those assemblies constituted the whole watch. Whenever Hora had to put down a partly completed watch to answer the phone, he lost only a small part of his work. So he made his watches much faster and more efficiently than did Tempus.

Complex systems can evolve from simple systems only if there are stable intermediate forms. The resulting complex forms will naturally be hierarchic. That may explain why hierarchies are so common in the systems nature presents to us. Among all possible complex forms, hierarchies are the only ones that have had the time to evolve.[5]

Hierarchies are brilliant systems inventions, not only because they give a system stability and resilience, but also because they reduce the amount of information that any part of the system has to keep track of.

In hierarchical systems relationships *within* each subsystem are denser and stronger than relationships *between* subsystems. Everything is still connected to everything else, but not equally strongly. People in the same university department talk to each other more than they talk to people in other departments. The cells that constitute the liver are in closer communication with each other than they are with the cells of the heart. If these differential information links within and between each level of the hierarchy are designed right, feedback delays are minimized. No level is overwhelmed with information. The system works with efficiency and resilience.

Hierarchical systems are partially decomposable. They can be taken apart and the subsystems with their especially dense information links can function, at least partially, as systems in their own right. When hierarchies break down, they usually split along their subsystem boundaries. Much can be learned by taking apart systems at different hierarchical levels—cells or organs, for example—and studying them separately. Hence, systems thinkers would say, the reductionist dissection of regular science teaches us a lot. However, one should not lose sight of the important relationships that bind each subsystem to the others and to the higher levels of the hierarchy, or one will be in for surprises.

If you have a liver disease, for example, a doctor usually can treat it without paying much attention to your heart or your tonsils (to stay on the same hierarchical level) or your personality (to move up a level or two) or the DNA in the nuclei of the liver cells (to move down several levels). There are just enough exceptions to that rule, however, to reinforce the necessity of stepping back to consider the whole hierarchy. Maybe your job exposes you to a chemical that is damaging your liver. Maybe the disease originates in a malfunction of the DNA.

What you need to think about may change over time, as self-organizing systems evolve new degrees of hierarchy and integration. The energy systems of nations were once almost completely decomposable one from another. That is no longer true. People whose thinking has not evolved as fast as the energy economy has may be shocked to discover how dependent they have become on resources and decisions halfway around the world.

You can watch self-organizing systems form hierarchies. A self-employed person gets too much work and hires some helpers. A small, informal nonprofit organization attracts many members and a bigger budget and one day the members decide, "Hey, we need someone to organize all this." A cluster of dividing cells differentiates into special functions and generates a branching circulatory system to feed all cells, and a branching nervous system to coordinate them.

Hierarchies evolve from the lowest level up—from the pieces to the whole, from cell to organ to organism, from individual to team, from actual production to management of production. Early farmers decided to come together and form cities for self-protection and for making trade more efficient. Life started with single-cell bacteria, not with elephants. The original purpose of a hierarchy is always to help its originating subsystems do their jobs better. This is something, unfortunately, that both the higher and the lower levels of a greatly articulated hierarchy easily can forget. Therefore, many systems are not meeting our goals because of malfunctioning hierarchies.

If a team member is more interested in personal glory than in the team winning, he or she can cause the team to lose. If a body cell breaks free from its hierarchical function and starts multiplying wildly, we call it a cancer. If students think their purpose is to maximize personal grades instead of seeking knowledge, cheating and other counterproductive behaviors break out. If a single corporation bribes the government to favor that corporation, the advantages of the competitive market and the good of the whole society are eroded.

When a subsystem's goals dominate at the expense of the total system's goals, the resulting behavior is called **suboptimization.**

Just as damaging as suboptimization, of course, is the problem of too much central control. If the brain controlled each cell so tightly that the cell could not perform its self-maintenance functions, the whole organism could die. If central rules and regulations prevent students or faculty from exploring fields of knowledge freely, the purpose of the university is not served. The coach of a team might interfere with the on-the-spot perceptions of a good player, to the detriment of the team. Economic examples of overcontrol from the top, from companies to nations, are the causes of some of the great catastrophes of history, all of which are by no means behind us.

To be a highly functional system, hierarchy must balance the welfare, freedoms, and responsibilities of the subsystems and total system—there must be enough central control to achieve coordination toward the large-system goal, and enough autonomy to keep all subsystems flourishing, functioning, and self-organizing.

Resilience, self-organization, and hierarchy are three of the reasons dynamic systems can work so well. Promoting or managing for these properties of a system can improve its ability to function well over the long term— to be sustainable. But watching how systems behave also can be full of surprises.

> Hierarchical systems evolve from the bottom up. The purpose of the upper layers of the hierarchy is to serve the purposes of the lower layers.

FOUR

Why Systems Surprise Us

The trouble . . . is that we are terrifyingly ignorant. The most learned of us are ignorant. . . . The acquisition of knowledge always involves the revelation of ignorance—almost *is* the revelation of ignorance. Our knowledge of the world instructs us first of all that the world is greater than our knowledge of it.

—Wendell Berry,[1] writer and Kentucky farmer

The simple systems in the zoo may have perplexed you with their behavior. They continue to surprise me, although I have been teaching them for years. That you and I are surprised says as much about us as it does about dynamic systems. The interactions between what I think I know about dynamic systems and my experience of the real world never fails to be humbling. They keep reminding me of three truths:

1. Everything we think we know about the world is a model. Every word and every language is a model. All maps and statistics, books and databases, equations and computer programs are models. So are the ways I picture the world in my head—my *mental* models. None of these is or ever will be the *real* world.
2. Our models usually have a strong congruence with the world. That is why we are such a successful species in the biosphere. Especially complex and sophisticated are the mental models we develop from direct, intimate experience of nature, people, and organizations immediately around us.
3. However, and conversely, our models fall far short of representing the world fully. That is why we make mistakes and why we are regularly surprised. In our heads, we can keep track of only a few variables at

one time. We often draw illogical conclusions from accurate assumptions, or logical conclusions from inaccurate assumptions. Most of us, for instance, are surprised by the amount of growth an exponential process can generate. Few of us can intuit how to damp oscillations in a complex system.

In short, this book is poised on a duality. We know a tremendous amount about how the world works, but not nearly enough. Our knowledge is amazing; our ignorance even more so. We can improve our understanding, but we can't make it perfect. I believe both sides of this duality, because I have learned much from the study of systems.

> **Everything we think we know about the world is a model.** Our models do have a strong congruence with the world. Our models fall far short of representing the real world fully.

This chapter describes some of the reasons why dynamic systems are so often surprising. Alternately, it is a compilation of some of the ways our mental models fail to take into account the complications of the real world—at least those ways that one can see from a systems perspective. It is a warning list. Here is where hidden snags lie. You can't navigate well in an interconnected, feedback-dominated world unless you take your eyes off short-term events and look for long-term behavior and structure; unless you are aware of false boundaries and bounded rationality; unless you take into account limiting factors, nonlinearities and delays. You are likely to mistreat, misdesign, or misread systems if you don't respect their properties of resilience, self-organization, and hierarchy.

The bad news, or the good news, depending on your need to control the world and your willingness to be delighted by its surprises, is that even if you do understand all these system characteristics, you may be surprised less often, but you will still be surprised.

Beguiling Events

A system is a big black box
Of which we can't unlock the locks,
And all we can find out about
Is what goes in and what comes out.

Perceiving input-output pairs,
Related by parameters,
Permits us, sometimes, to relate
An input, output and a state.
If this relation's good and stable
Then to predict we may be able,
But if this fails us—heaven forbid!
We'll be compelled to force the lid!

—Kenneth Boulding,[2] economist

Systems fool us by presenting themselves—or we fool ourselves by seeing the world—as a series of events. The daily news tells of elections, battles, political agreements, disasters, stock market booms or busts. Much of our ordinary conversation is about specific happenings at specific times and places. A team wins. A river floods. The Dow Jones Industrial Average hits 10,000. Oil is discovered. A forest is cut. Events are the outputs, moment by moment, from the black box of the system.

Events can be spectacular: crashes, assassinations, great victories, terrible tragedies. They hook our emotions. Although we've seen many thousands of them on our TV screens or the front page of the paper, each one is different enough from the last to keep us fascinated (just as we never lose our fascination with the chaotic twists and turns of the weather). It's endlessly engrossing to take in the world as a series of events, and constantly surprising, because that way of seeing the world has almost no predictive or explanatory value. Like the tip of an iceberg rising above the water, events are the most visible aspect of a larger complex—but not always the most important.

We are less likely to be surprised if we can see how events accumulate into dynamic patterns of *behavior*. The team is on a winning streak. The variance of the river is increasing, with higher floodwaters during rains and lower flows during droughts. The Dow has been trending up for two years. Discoveries of oil are becoming less frequent. The felling of forests is happening at an ever-increasing rate.

The behavior of a system is its performance over time—its growth, stagnation, decline, oscillation, randomness, or evolution. If the news did a better job of putting events into historical context, we would have better behavior-level understanding, which is deeper than event-level understanding. When a systems thinker encounters a problem, the first thing he or she does is look for data, time graphs, the history of the system. That's because long-term behavior provides clues to the underlying system structure. And structure is the key to understanding not just *what* is happening, but *why*.

The structure of a system is its interlocking stocks, flows, and feedback loops. The diagrams with boxes and arrows (my students call them "spaghetti-and-meatball diagrams") are pictures of system structure. Structure determines what behaviors are latent in the system. A goal-seeking balancing feedback loop approaches or holds a dynamic equilibrium. A reinforcing feedback loop generates exponential growth. The two of them linked together are capable of growth, decay, or equilibrium. If they also contain delays, they may produce oscillations. If they work in periodic, rapid bursts, they may produce even more surprising behaviors.

> System structure is the source of system behavior. System behavior reveals itself as a series of events over time.

Systems thinking goes back and forth constantly between structure (diagrams of stocks, flows, and feedback) and behavior (time graphs). Systems thinkers strive to understand the connections between the hand releasing the Slinky (event) and the resulting oscillations (behavior) and the mechanical characteristics of the Slinky's helical coil (structure).

Simple examples like a Slinky make this event-behavior-structure distinction seem obvious. In fact, much analysis in the world goes no deeper than events. Listen to every night's explanation of why the stock market did what it did. Stocks went up (down) because the U.S. dollar fell (rose), or the prime interest rate rose (fell), or the Democrats won (lost), or one country invaded another (or didn't). Event-event analysis.

These explanations give you no ability to predict what will happen tomorrow. They give you no ability to change the behavior of the system—to make the stock market less volatile or a more reliable indicator of the health of corporations or a better vehicle to encourage investment, for instance.

Most economic analysis goes one level deeper, to behavior over time. Econometric models strive to find the statistical links among past trends in income, savings, investment, government spending, interest rates, output, or whatever, often in complicated equations.

These behavior-based models are more useful than event-based ones, but they still have fundamental problems. First, they typically overemphasize system flows and underemphasize stocks. Economists follow the behavior of flows, because that's where the interesting variations and most rapid changes in systems show up. Economic news reports on the national production (flow) of goods and services, the GNP, rather than the total physical capital (stock) of the nation's factories and farms and businesses that produce those goods and services. But without seeing how stocks affect their related flows through feedback processes, one cannot understand the dynamics of economic systems or the reasons for their behavior.

Second, and more seriously, in trying to find statistical links that relate flows to each other, econometricians are searching for something that does not exist. There's no reason to expect any flow to bear a stable relationship to any other flow. Flows go up and down, on and off, in all sorts of combinations, in response to stocks, not to other flows.

Let me use a simple example to explain what I mean. Suppose you knew nothing at all about thermostats, but you had a lot of data about past heat flows into and out of the room. You could find an equation telling you how those flows have varied together in the past, because under ordinary circumstances, being governed by the same stock (temperature of the room), they do vary together.

Your equation would hold, however, only until something changes in the system's structure—someone opens a window or improves the insulation, or tunes the furnace, or forgets to order oil. You could predict tomorrow's room temperature with your equation, as long as the system didn't change or break down. But if you were asked to make the room warmer, or if the room temperature suddenly started plummeting and you had to fix it, or if you wanted to produce the same room temperature with a lower fuel bill, your behavior-level analysis wouldn't help you. You would have to dig into the system's structure.

That's why behavior-based econometric models are pretty good at predicting the near-term performance of the economy, quite bad at predicting the longer-term performance, and terrible at telling one how to improve the performance of the economy.

And that's one reason why systems of all kinds surprise us. We are too fascinated by the events they generate. We pay too little attention to their history. And we are insufficiently skilled at seeing in their history clues to the structures from which behavior and events flow.

Linear Minds in a Nonlinear World

Linear relationships are easy to think about: the more the merrier. Linear equations are solvable, which makes them suitable for textbooks. Linear systems have an important modular virtue: you can take them apart and put them together again—the pieces add up.

Nonlinear systems generally cannot be solved and cannot be added together. . . . Nonlinearity means that the act of playing the game has a way of changing the rules. . . . That twisted changeability makes nonlinearity hard to calculate, but it also creates rich kinds of behavior that never occur in linear systems.

—James Gleick, author of *Chaos: Making a New Science*[3]

We often are not very skilled in understanding the nature of relationships. A **linear relationship** between two elements in a system can be drawn on a graph with a straight line. It's a relationship with constant proportions. If I put 10 pounds of fertilizer on my field, my yield will go up by 2 bushels. If I put on 20 pounds, my yield will go up by 4 bushels. If I put on 30 pounds, I'll get an increase of 6 bushels.

A **nonlinear relationship** is one in which the cause does not produce a proportional effect. The relationship between cause and effect can only be drawn with curves or wiggles, not with a straight line. If I put 100 pounds of fertilizer on, my yield will go up by 10 bushels; if I put on 200, my yield will not go up at all; if I put on 300, my yield will go down. Why? I've damaged my soil with "too much of a good thing."

The world is full of nonlinearities.

So the world often surprises our linear-thinking minds. If we've learned that a small push produces a small response, we think that twice as big a push will produce twice as big a response. But in a nonlinear system, twice the push could produce one-sixth the response, or the response squared, or no response at all.

Here are some examples of nonlinearities:

- As the flow of traffic on a highway increases, car speed is affected only slightly over a large range of car density. Eventually, however, small further increases in density produce a rapid drop-off in speed. And when the number of cars on the highway builds up to a certain point, it can result in a traffic jam, and car speed drops to zero.
- Soil erosion can proceed for a long time without much affect on crop yield—until the topsoil is worn down to the depth of the root zone of the crop. Beyond that point, a little further erosion can cause yields to plummet.
- A little tasteful advertising can awaken interest in a product. A lot of blatant advertising can cause disgust for the product.

You can see why nonlinearities produce surprises. They foil the reasonable expectation that if a little of some cure did a little good, then a lot of it will do a lot of good—or alternately that if a little destructive action caused only a tolerable amount of harm, then more of that same kind of destruction will cause only a bit more harm. Reasonable expectations like these in a nonlinear world produce classic mistakes.

Nonlinearities are important not only because they confound our expectations about the relationship between action and response. They are even more important because they *change the relative strengths of feedback loops*. They can flip a system from one mode of behavior to another.

Nonlinearities are the chief cause of the shifting dominance that characterizes several of the systems in the zoo—the sudden swing between exponential growth caused by a dominant reinforcing loop, say, and then decline caused by a suddenly dominant balancing loop.

To take a dramatic example of the effects of nonlinearities, consider the destructive irruptions of the spruce budworm in North American forests.

INTERLUDE • *Spruce Budworms, Firs, and Pesticides*

Tree ring records show that the spruce budworm has been killing spruce and fir trees periodically in North America for at least 400 years. Until this century, no one much cared. The valuable tree for the lumber industry was the white pine. Spruce and fir were considered "weed species." Eventually, however, the stands of virgin pine were gone, and the lumber industry turned to spruce and fir. Suddenly the budworm was seen as a serious pest.

So, beginning in the 1950s, northern forests were sprayed with DDT to control the spruce budworm. In spite of the spraying, every year there was a budworm resurgence. Annual sprays were continued through the 1950s, 1960s, and 1970s, until DDT was banned. Then the sprays were changed to fenitrothion, acephate, Sevin, and methoxychlor.

Insecticides were no longer thought to be the ultimate answer to the budworm problem, but they were still seen as essential. "Insecticides buy time," said one forester, "That's all the forest manager wants; to preserve the trees until the mill is ready for them."

By 1980, spraying costs were getting unmanageable—the Canadian province of New Brunswick spent $12.5 million on budworm "control" that year. Concerned citizens were objecting to the drenching of the landscape with poisons. And, in spite of the sprays, the budworm was still killing as many as 20 million hectares (50 million acres) of trees per year.

C. S. Holling of the University of British Columbia and Gordon Baskerville of the University of New Brunswick put together a computer model to get a whole-system look at the budworm problem. They discovered that before the spraying began, the budworm had been barely detectable in most years. It was controlled by a number of predators, including birds, a spider, a parasitic wasp, and several diseases. Every few decades, however, there was a budworm outbreak, lasting from six to ten years. Then the budworm population would subside, eventually to explode again

The budworm preferentially attacks balsam fir, secondarily spruce. Balsam fir is the most competitive tree in the northern forest. Left to its own devices, it would crowd out spruce and birch, and the forest would become a monoculture of nothing but fir. Each budworm outbreak cuts back the fir population, opening the forest for spruce and birch. Eventually fir moves back in.

As the fir population builds up, the probability of an outbreak increases—*nonlinearly*. The reproductive potential of the budworm increases more than proportionately to the availability of its favorite food supply. The final trigger is two or three warm, dry springs, perfect for the survival of budworm larvae. (If you're doing event-level analysis, you will blame the outburst on the warm, dry springs.)

The budworm population grows too great for its natural enemies to hold in check—*nonlinearly*. Over a wide range of conditions, greater budworm populations result in more rapid multiplication of budworm predators. But beyond some point, the predators can multiply no faster. What was a reinforcing relationship—more budworms, faster predator multiplication—becomes a nonrelationship—more budworms, no faster predator multiplication—and the budworms take off, unimpeded.

Now only one thing can stop the outbreak: the insect reducing its own food supply by killing off fir trees. When that finally happens, the budworm population crashes—*nonlinearly*. The reinforcing loop of budworm reproduction yields dominance to the balancing loop of budworm starvation. Spruce and birch move into the spaces where the firs used to be, and the cycle begins again.

The budworm/spruce/fir system oscillates over decades, but it is ecologically stable within bounds. It can go on forever. The main effect of the budworm is to allow tree species other than fir to persist. But in this case what is ecologically stable is economically unstable. In eastern Canada, the economy is almost completely dependent on the logging industry, which is dependent on a steady supply of fir and spruce.

Many relationships in systems are nonlinear. Their relative strengths shift in disproportionate amounts as the stocks in the system shift. Nonlinearities in feedback systems produce shifting dominance of loops and many complexities in system behavior.

When industry sprays insecticides, it shifts the whole system to balance uneasily on different points within its nonlinear relationships. It kills off not only the pest, but the natural enemies of the pest, thereby weakening the feedback loop that normally keeps the budworms in check. It keeps the density of fir high, moving the budworms up their nonlinear reproduction curve to the point at which they're perpetually on the edge of population explosion.

The forest management practices have set up what Holling calls "persistent semi-outbreak conditions" over larger and larger areas. The managers have found themselves locked into a policy in which there is an incipient volcano bubbling, such that, if the policy fails, there will be an outbreak of an intensity that has never been seen before."[4]

Nonexistent Boundaries

When we think in terms of systems, we see that a fundamental misconception is embedded in the popular term "side-effects.". . . This phrase means roughly "effects which I hadn't foreseen or don't want to think about.". . . Side-effects no more deserve the adjective "side" than does the "principal" effect. It is hard to think in terms of systems, and we eagerly warp our language to protect ourselves from the necessity of doing so.

—Garrett Hardin,[5] ecologist

Remember the clouds in the structural diagrams of Chapters One and Two? Beware of clouds! They are prime sources of system surprises.

Clouds stand for the beginnings and ends of flows. They are stocks—sources and sinks—that are being ignored at the moment for the purposes of simplifying the present discussion. They mark the boundary of the system diagram. They rarely mark a real boundary, because systems rarely have real boundaries. Everything, as they say, is connected to everything else, and not neatly. There is no clearly determinable boundary between the sea and the land, between sociology and anthropology, between an automobile's exhaust and your nose. There are only boundaries of word, thought, perception, and social agreement—artificial, mental-model boundaries.

The greatest complexities arise exactly at boundaries. There are Czechs on the German side of the border and Germans on the Czech side of the border. Forest species extend beyond the edge of the forest into the field; field species penetrate partway into the forest. Disorderly, mixed-up borders are sources of diversity and creativity.

In our system zoo, for instance, I showed the flow of cars into a car dealer's inventory as coming from a cloud. Of course, cars don't come from a cloud, they come from the transformation of a stock of raw materials, with the help of capital, labor, energy, technology, and management (the means of production). Similarly, the flow of cars out of the inventory goes not to a cloud, but through sales to the households or businesses of consumers.

Whether it is important to keep track of raw materials or consumers' home stocks (whether it is legitimate to replace them in a diagram with clouds) depends on whether these stocks are likely to have a significant influence on

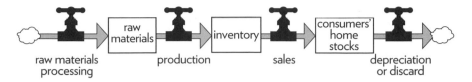

raw materials
processing   production   sales   depreciation
                                                       or discard

Figure 47. Revealing some of the stocks behind the clouds.

the behavior of the system over the time period of interest. If raw materials are guaranteed to be abundant and consumers continue to demand the products, then clouds will do. But if there could be a materials shortage or a product glut, and if we drew a mental boundary around the system that did not include these stocks, then we could be surprised by future events.

There are still clouds in Figure 47. The boundary can be expanded further. Processed raw materials come from chemical plants, smelters, or refineries, whose input comes, ultimately, from the earth. Processing creates not only products, but also employment, wages, profits, and pollution. Discarded consumers' stocks go to landfills or incinerators or recycling centers, from which they go on to have further effects on society and the environment. Landfills leach into drinking-water wells, incinerators produce smoke and ash, recycling centers move materials back into the production stream.

Whether it's important to think about the full flow from mine to dump, or as industry calls it, "from cradle to grave," depends on who wants to know, for what purpose, over how long. In the long term, the full flow is important and, as the physical economy grows and society's "ecological footprint" expands, the long term is increasingly coming to be the short term. Landfills fill up with a suddenness that has been surprising for people whose mental models picture garbage as going "away," into some sort of a cloud. Sources of raw materials—mines, wells, and oil fields—can be exhausted with surprising suddenness too.

With a long enough time horizon, even mines and dumps are not the end of the story. The great geological cycles of the earth keep moving materials around, opening and closing seas, raising up and wearing down mountains. Eons from now, everything put in a dump will end up on the top of a mountain or deep under the sea. New deposits of metals and fuels will form. On planet Earth there are no system "clouds," no ultimate boundaries. Even real clouds in the sky are part of a hydrological cycle. Everything physical comes from somewhere, everything goes somewhere, everything keeps moving.

Which is not to say that every model, mental or computer, has to follow each connection until it includes the whole planet. Clouds are a necessary part of models that describe metaphysical flows. Anger literally "comes out of a cloud," as does love, hatred, self esteem, and so on. If we're to understand anything, we have to simplify, which means we have to make boundaries. Often that's a safe thing to do. It's usually not a problem, for example, to think of populations with births and deaths coming from and going to clouds, as in Figure 48.

Figure 48. More clouds.

Figure 48 shows actual "cradle to grave" boundaries. Even these boundaries would be unserviceable, however, if the population in question experienced significant in- or out-migration, or if the problem under discussion was limited cemetery space.

The lesson of boundaries is hard even for systems thinkers to get. There is no single, legitimate boundary to draw around a system. We have to invent boundaries for clarity and sanity; and boundaries can produce problems when we forget that we've artificially created them.

> There are no separate systems. The world is a continuum. Where to draw a boundary around a system depends on the purpose of the discussion—the questions we want to ask.

When you draw boundaries too narrowly, the system surprises you. For example, if you try to deal with urban traffic problems without thinking about settlement patterns, you build highways, which attract housing developments along their whole length. Those households, in turn, put more cars on the highways, which then become just as clogged as before.

If you try to solve a sewage problem by throwing the waste into a river, the towns downstream make it clear that the boundary for thinking about sewage has to include the whole river. It might also have to include the soil and groundwater surrounding the river. It probably doesn't have to include the next watershed or the planetary hydrological cycle.

Planning for a national park used to stop at the physical boundary of the park. But park boundaries around the world are regularly crossed by nomadic peoples, by migrating wildlife, by waters that flow into, out of, or under the park, by the effects of economic development at the park's edges, by acid rain, and now by a climate changing from greenhouse gases in the atmosphere. Even without climate change, to manage a park you have to think about a boundary wider than the official perimeter.

Systems analysts often fall into the opposite trap: making boundaries too large. They have a habit of producing diagrams that cover several pages with small print and many arrows connecting everything with everything. *There* is the system! they say. If you have considered anything less, you are academically illegitimate.

This "my model is bigger than your model" game results in enormously complicated analyses, which produce piles of information that may only serve to obscure the answers to the questions at hand. For example, modeling the earth's climate in full detail is interesting for many reasons, but may not be necessary for figuring out how to reduce a country's CO_2 emissions to reduce climate change.

The right boundary for thinking about a problem rarely coincides with the boundary of an academic discipline, or with a political boundary. Rivers make handy borders between countries, but the worst possible borders for managing the quantity and quality of the water. Air is worse than water in its insistence on crossing political borders. National boundaries mean nothing when it comes to ozone depletion in the stratosphere, or greenhouse gases in the atmosphere, or ocean dumping.

Ideally, we would have the mental flexibility to find the appropriate boundary for thinking about each new problem. We are rarely that flexible. We get attached to the boundaries our minds happen to be accustomed to. Think how many arguments have to do with boundaries—national boundaries, trade boundaries, ethnic boundaries, boundaries between public and private responsibility, and boundaries between the rich and the poor, polluters and pollutees, people alive now and people who will come in the future. Universities can maintain disputes for years about the boundaries between economics and government, art and art history, literature and literary criticism. Too often, universities are living monuments to boundary rigidity.

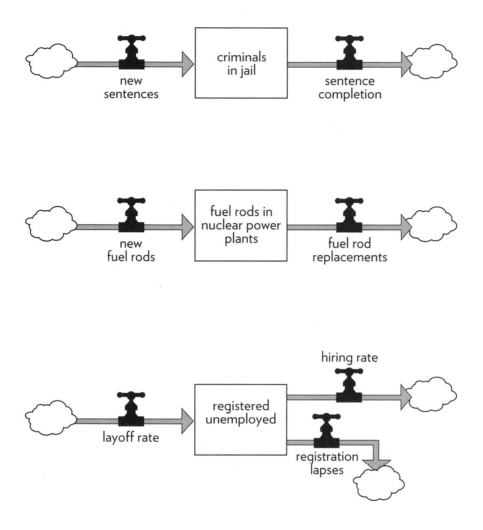

Figure 49. Examples of more clouds. These are systems in which a boundary or cloud should not stop you from thinking beyond the borders of the system, but start you thinking beyond those borders. What is driving the supply of people being given new sentences? Where do the fuel rods go after replacement? What happens to an unemployed person whose registration for unemployment lapses?

It's a great art to remember that *boundaries are of our own making, and that they can and should be reconsidered for each new discussion, problem, or purpose.* It's a challenge to stay creative enough to drop the boundaries that worked for the last problem and to find the most appropriate set of boundaries for the next question. It's also a necessity, if problems are to be solved well.

Layers of Limits

Systems surprise us because our minds like to think about single causes neatly producing single effects. We like to think about one or at most a few things at a time. And we don't like, especially when our own plans and desires are involved, to think about limits.

But we live in a world in which many causes routinely come together to produce many effects. Multiple inputs produce multiple outputs, and virtually all of the inputs, and therefore outputs, are limited. For example, an industrial manufacturing process needs:

- capital
- labor
- energy
- raw materials
- land
- water
- technology
- credit
- insurance
- customers
- good management
- public-funded infrastructure and government services (such as police and fire protection and education for managers and workers)
- functioning families to bring up and care for both producers and consumers
- a healthy ecosystem to supply or support all these inputs and to absorb or carry away their wastes

A patch of growing grain needs:

- sunlight
- air
- water
- nitrogen
- phosphorus

- potassium
- dozens of minor nutrients
- a friable soil and the services of a microbial soil community
- some system to control weeds and pests
- protection from the wastes of the industrial manufacturer

It was with regard to grain that Justus von Liebig came up with his famous "law of the minimum." It doesn't matter how much nitrogen is available to the grain, he said, if what's short is phosphorus. It does no good to pour on more phosphorus, if the problem is low potassium.

Bread will not rise without yeast, no matter how much flour it has. Children will not thrive without protein, no matter how many carbohydrates they eat. Companies can't keep going without energy, no matter how many customers they have—or without customers, no matter how much energy they have.

This concept of a **limiting factor** is simple and widely misunderstood. Agronomists assume, for example, that they know what to put in artificial fertilizer, because they have identified many of the major and minor nutrients in good soil. Are there any essential nutrients they have not identified? How do artificial fertilizers affect soil microbe communities? Do they interfere with, and therefore limit, any other functions of good soil? And what limits the production of artificial fertilizers?

> At any given time, the input that is most important to a system is the one that is most limiting.

Rich countries transfer capital or technology to poor ones and wonder why the economies of the receiving countries still don't develop, never thinking that capital or technology may not be the most limiting factors.

Economics evolved in a time when labor and capital were the most common limiting factors to production. Therefore, most economic production functions keep track only of these two factors (and sometimes technology). As the economy grows relative to the ecosystem, however, and the limiting factors shift to clean water, clean air, dump space, and acceptable forms of energy and raw materials, the traditional focus on only capital and labor becomes increasingly unhelpful.

One of the classic models taught to systems students at MIT is Jay Forrester's corporate-growth model. It starts with a successful young company, growing rapidly. The problem for this company is to recognize and deal with its shifting limits—limits that change in response to the company's own growth.

The company may hire salespeople, for example, who are so good that they generate orders faster than the factory can produce. Delivery delays increase and customers are lost, because production capacity is the most limiting factor. So the managers expand the capital stock of production plants. New people are hired in a hurry and trained too little. Quality suffers and customers are lost because labor skill is the most limiting factor. So management invests in worker training. Quality improves, new orders pour in, and the order-fulfillment and record-keeping system clogs. And so forth.

There are layers of limits around every growing plant, child, epidemic, new product, technological advance, company, city, economy, and population. Insight comes not only from recognizing which factor is limiting, but from seeing that *growth itself depletes or enhances limits* and therefore changes what is limiting. The interplay between a growing plant and the soil, a growing company and its market, a growing economy and its resource base, is dynamic. Whenever one factor ceases to be limiting, growth occurs, and the growth itself changes the relative scarcity of factors until another becomes limiting. To shift attention from the abundant factors to the next potential limiting factor is to gain real understanding of, and control over, the growth process.

Any physical entity with multiple inputs and outputs—a population, a production process, an economy—is surrounded by layers of limits. As the system develops, it interacts with and affects its own limits. The growing entity and its limited environment together form a coevolving dynamic system.

Understanding layers of limits and keeping an eye on the next upcoming limiting factor is not a recipe for perpetual growth, however. For any physical entity in a finite environment, perpetual growth is impossible. Ultimately, the choice is not to grow forever but to decide what limits to live within. If a company produces a perfect product or service at an affordable price, it will be swamped with orders until it grows to the point at which some limit decreases the perfection of the product or raises its price. If a city meets the needs of all its inhabitants better than any other city, people will flock there until some limit brings down the city's ability to satisfy peoples' needs.[6]

> **Any physical entity with multiple inputs and outputs is surrounded by layers of limits.**

There always will be limits to growth. They can be self-imposed. If they aren't, they will be system-imposed. No physical entity can grow forever. If company managers, city governments, the human population do not choose and enforce their own limits to keep growth within the capacity of the supporting environment, then the environment will choose and enforce limits.

> **There always will be limits to growth. They can be self-imposed. If they aren't, they will be system-imposed.**

Ubiquitous Delays

I realize with fright that my impatience for the re-establishment of democracy had something almost communist in it; or, more generally, something rationalist. I had wanted to make history move ahead in the same way that a child pulls on a plant to make it grow more quickly.

I believe we must learn to wait as we learn to create. We have to patiently sow the seeds, assiduously water the earth where they are sown and give the plants the time that is their own. One cannot fool a plant any more than one can fool history.

—Václav Havel,[7] playwright, last President of Czechoslovakia and first president of the Czech Republic

It takes time for a plant or a forest or a democracy to grow; time for letters put into a mailbox to reach their destinations; time for consumers to absorb information about changing prices and alter their buying behavior, or for a nuclear power plant to be built, or a machine to wear out, or a new technology to penetrate an economy.

We are surprised over and over again at how much time things take. Jay Forrester used to tell us, when we were modeling a construction or processing delay, to ask everyone in the system how long they thought the delay was, make our best guess, and then multiply by three. (That correction factor also works perfectly, I have found, for estimating how long it will take to write a book!)

Delays are ubiquitous in systems. Every stock is a delay. Most flows have delays—shipping delays, perception delays, processing delays, maturation delays. Here are just a few of the delays we have found important to include in various models we have made:

- The delay between catching an infectious disease and getting sick enough to be diagnosed—days to years, depending on the disease.
- The delay between pollution emission and the diffusion or percolation or concentration of the pollutant in the ecosystem to the point at which it does harm.
- The gestation and maturation delay in building up breeding populations of animals or plants, causing the characteristic oscillations of commodity prices: 4-year cycles for pigs, 7 years for cows, 11 years for cocoa trees.[8]
- The delay in changing the social norms for desirable family size—at least one generation.
- The delay in retooling a production stream and the delay in turning over a capital stock. It takes 3 to 8 years to design a new car and bring it to the market. That model may have 5 years of life on the new-car market. Cars stay on the road an average of 10 to 15 years.

Just as the appropriate boundaries to draw around one's picture of a system depend on the purpose of the discussion, so do the important delays. If you're worrying about oscillations that take weeks, you probably don't have to think about delays that take minutes, or years. If you're concerned about the decades-long development of a population and economy, you usually can ignore oscillations that take weeks. The world peeps, squawks, bangs, and thunders at many frequencies all at once. What is a significant delay depends—usually—on which set of frequencies you're trying to understand.

The systems zoo has already demonstrated how important delays in feedback are to the behavior of systems. Changing the length of a delay may utterly change behavior. Delays are often sensitive leverage points for policy, if they can be made shorter or longer. You can see why that is. If a decision point in a system (or a person working in that part of the system) is responding to delayed information, or responding with a delay, the decisions will be off target. Actions will be too much or too little to achieve the decision maker's goals. On the other hand, if action is taken too fast, it may nervously amplify short-term variation and create unnecessary instability. Delays determine how fast systems can react, how accurately they hit their targets, and how timely is the information passed around a system. Overshoots, oscillations, and collapses are always caused by delays.

Understanding delays helps one understand why Mikhail Gorbachev could transform the information system of the Soviet Union virtually overnight, but not the physical economy. (That takes decades.) It helps one see why the absorption of East Germany by West Germany produced more hardship over a longer time than the politicians foresaw. Because of long delays in building new power plants, the electricity industry is plagued with cycles of overcapacity and then undercapacity leading to brownouts. Because of decades-long delays as the earth's oceans respond to warmer temperatures, human fossil-fuel emissions have already induced changes in climate that will not be fully revealed for a generation or two.

> When there are long delays in feedback loops, some sort of foresight is essential. To act only when a problem becomes obvious is to miss an important opportunity to solve the problem.

Bounded Rationality

> As every individual, therefore, endeavours as much as he can both to employ his capital in the support of domestic industry, and so to direct that industry that its produce may be of greatest value... he generally, indeed, neither intends to promote the public interest, nor knows how much he is promoting it. . . . He intends his own security; . . . he intends only his own gain and he is in this . . . led by an invisible hand to promote an end which was no part of his intention. By pursuing his own interest he frequently promotes that of society more effectually than when he really intends to promote it.
>
> Adam Smith,[9] 18th century political economist

It would be so nice if the "invisible hand" of the market really did lead individuals to make decisions that add up to the good of the whole. Then not only would material selfishness be a social virtue, but mathematical models of the economy would be much easier to make. There would be no need to think about the good of other people or about the operations of complex feedback systems. No wonder Adam Smith's model has had such strong appeal for two hundred years!

Unfortunately, the world presents us with multiple examples of people acting rationally in their short-term best interests and producing aggregate results that no one likes. Tourists flock to places like Waikiki or Zermatt and then complain that those places have been ruined by all the tourists. Farmers produce surpluses of wheat, butter, or cheese, and prices plummet. Fishermen overfish and destroy their own livelihood. Corporations collectively make investment decisions that cause business-cycle downturns. Poor people have more babies than they can support.

Why?

Because of what World Bank economist Herman Daly calls the "invisible foot" or what Nobel Prize–winning economist Herbert Simon calls **bounded rationality**.[10]

Bounded rationality means that people make quite reasonable decisions based on the information they have. But they don't have perfect information, especially about more distant parts of the system. Fishermen don't know how many fish there are, much less how many fish will be caught by other fishermen that same day.

Businessmen don't know for sure what other businessmen are planning to invest, or what consumers will be willing to buy, or how their products will compete. They don't know their current market share, and they don't know the size of the market. Their information about these things is incomplete and delayed, and their own responses are delayed. So they systematically under- and overinvest.

We are not omniscient, rational optimizers, says Simon. Rather, we are blundering "satisficers," attempting to meet (*satisfy*) our needs well enough (*sufficiently*) before moving on to the next decision.[11] We do our best to further our own nearby interests in a rational way, but we can take into account only what we know. We don't know what others are planning to do, until they do it. We rarely see the full range of possibilities before us. We often don't foresee (or choose to ignore) the impacts of our actions on the whole system. So instead of finding a long-term optimum, we discover within our limited purview a choice we can live with for now, and we stick to it, changing our behavior only when forced to.

We don't even interpret perfectly the imperfect information that we do have, say behavioral scientists. We misperceive risk, assuming that some things are much more dangerous than they really are and others much less. We live in an exaggerated present—we pay too much attention to recent experience and too little attention to the past, focusing on current events rather than long-term behavior. We discount the future at rates that make no economic or ecological sense. We don't give all incoming signals their appropriate weights. We don't let in at all news we don't like, or information that doesn't fit our mental models. Which is to say, we don't even make decisions that optimize our own individual good, much less the good of the system as a whole.

When the theory of bounded rationality challenged two hundred years of economics based on the teachings of political economist Adam Smith, you can imagine the controversy that resulted—one that is far from over. Economic theory as derived from Adam Smith assumes first that *homo economicus* acts with perfect optimality on complete information, and second that when many of the species *homo economicus* do that, their actions add up to the best possible outcome for everybody.

Neither of these assumptions stands up long against the evidence. In the next chapter on system traps and opportunities, I will describe some of the most commonly encountered structures that can cause bounded rationality to lead to disaster. They include such familiar phenomena as addiction, policy resistance, arms races, drift to low performance, and the tragedy of the commons. For now, I want to make just one point about the biggest surprise that comes from not understanding bounded rationality.

Suppose you are for some reason lifted out of your accustomed place in society and put in the place of someone whose behavior you have never understood. Having been a staunch critic of government, you suddenly become part of government. Or having been a laborer in opposition to management, you become management (or vice versa). Perhaps having been an environmental critic of big business, you find yourself making environmental decisions for big business. Would that such transitions could happen much more often, in all directions, to broaden everyone's horizons!

In your new position, you experience the information flows, the incentives and disincentives, the goals and discrepancies, the pressures—the bounded rationality—that goes with that position. It's possible that you could retain your memory of how things look from another angle, and that you burst forth with innovations that transform the system, but it's distinctly unlikely. If you become a manager, you probably will stop seeing labor as a deserving partner in production, and start seeing it as a cost to be minimized. If you become a financier, you probably will overinvest during booms and underinvest during busts, along with all the other financiers. If you become very poor, you will see the short-term rationality, the hope, the opportunity, the necessity of having many children. If you are now a fisherman with a mortgage on your boat, a family to support, and imperfect knowledge of the state of the fish population, you will overfish.

We teach this point by playing games in which students are put into situations in which they experience the realistic, partial information streams seen by various actors in real systems. As simulated fishermen, they overfish. As ministers of simulated developing nations, they favor the needs of their industries over the needs of their people. As the upper class, they feather their own nests; as the lower class, they become apathetic or rebellious. So would you. In the famous Stanford prison experiment by psychologist Philip Zimbardo, players even took on, in an amazingly short time, the attitudes and behaviors of prison guards and prisoners.[12]

Seeing how individual decisions are rational within the bounds of the information available does not provide an excuse for narrow-minded behavior. It provides an understanding of why that behavior arises. Within the bounds of what a person in that part of the system can see and know, the behavior is reasonable. Taking out one individual from a position of bounded rationality and putting in another person is not likely to make much difference. Blaming the individual rarely helps create a more desirable outcome.

Change comes first from stepping outside the limited information that can be seen from any single place in the system and getting an overview. From a wider perspective, information flows, goals, incentives, and disincentives can be restructured so that separate, bounded, rational actions do add up to results that everyone desires.

It's amazing how quickly and easily behavior changes can come, with even slight enlargement of bounded rationality, by providing better, more complete, timelier information.

INTERLUDE • *Electric Meters in Dutch Houses*

Near Amsterdam, there is a suburb of single-family houses all built at the same time, all alike. Well, nearly alike. For unknown reasons it happened that some of the houses were built with the electric meter down in the basement. In other houses, the electric meter was installed in the front hall.

These were the sort of electric meters that have a glass bubble with a small horizontal metal wheel inside. As the household uses more electricity, the wheel turns faster and a dial adds up the accumulated kilowatt-hours.

During the oil embargo and energy crisis of the early 1970s, the Dutch began to pay close attention to their energy use. It was discovered that some of the houses in this subdivision used one-third less electricity than the other houses. No one could explain this. All houses were charged the same price for electricity, all contained similar families.

The difference, it turned out, was in the position of the electric meter. The families with high electricity use were the ones with the meter in the basement, where people rarely saw it. The ones with low use had the meter in the front hall where people passed, the little wheel turning around, adding up the monthly electricity bill many times a day.[13]

Some systems are structured to function well despite bounded rationality. The right feedback gets to the right place at the right time. Under ordinary circumstances, your liver gets just the information it needs to do its job. In undisturbed ecosystems and traditional cultures, the average individual, species, or population, left to its own devices, behaves in ways that serve and stabilize the whole. These systems and others are self-regulatory. They do not cause problems. We don't have government agencies and dozens of failed policies about them.

Since Adam Smith, it has been widely believed that the free, competitive market is one of these properly structured self-regulating systems. In some ways, it is. In other ways, obvious to anyone who is willing to look, it isn't. A free market does allow producers and consumers, who have the best information about production opportunities and consumption choices, to make fairly uninhibited and locally rational decisions. But those decisions can't, by themselves, correct the overall system's tendency to create monopolies and undesirable side effects (externalities), to discriminate against the poor, or to overshoot its sustainable carrying capacity.

To paraphrase a common prayer: God grant us the serenity to exercise our bounded rationality freely in the systems that are structured appropriately, the courage to restructure the systems that aren't, and the wisdom to know the difference!

The bounded rationality of each actor in a system—determined by the information, incentives, disincentives, goals, stresses, and constraints impinging on that actor— may or may not lead to decisions that further the welfare of the system as a whole. If they do not, putting new actors into the same system will not improve the system's performance. What makes a difference is redesigning the system to improve the information, incentives, disincentives, goals, stresses, and constraints that have an effect on specific actors.

> **The bounded rationality of each actor in a system may not lead to decisions that further the welfare of the system as a whole.**

FIVE

System Traps . . .
and Opportunities

Rational elites . . . know everything there is to know about their self-contained technical or scientific worlds, but lack a broader perspective. They range from Marxist cadres to Jesuits, from Harvard MBAs to army staff officers. . . . They have a common underlying concern: how to get their particular system to function. Meanwhile . . . civilization becomes increasingly directionless and incomprehensible.

—John Ralston Saul,[1] political scientist

Delays, nonlinearities, lack of firm boundaries, and other properties of systems that surprise us are found in just about any system. Generally, they are not properties that can or should be changed. The world is nonlinear. Trying to make it linear for our mathematical or administrative convenience is not usually a good idea even when feasible, and it is rarely feasible. Boundaries are problem-dependent, evanescent, and messy; they are also necessary for organization and clarity. Being less surprised by complex systems is mainly a matter of learning to expect, appreciate, and use the world's complexity.

But some systems are more than surprising. They are perverse. These are the systems that are structured in ways that produce truly problematic behavior; they cause us great trouble. There are many forms of systems trouble, some of them unique, but many strikingly common. We call the system structures that produce such common patterns of problematic behavior **archetypes**. Some of the behaviors these archetypes manifest are addiction, drift to low performance, and escalation. These are so prevalent that I had no problem finding in just one week of the *International Herald Tribune* enough examples to illustrate each of the archetypes described in this chapter.

Understanding archetypal problem-generating structures is not enough. Putting up with them is impossible. They need to be changed. The destruction they cause is often blamed on particular actors or events, although it is actually a consequence of system structure. Blaming, disciplining, firing, twisting policy levers harder, hoping for a more favorable sequence of driving events, tinkering at the margins—these standard responses will not fix structural problems. That is why I call these archetypes "traps."

But system traps can be escaped—by recognizing them in advance and not getting caught in them, or by altering the structure—by reformulating goals, by weakening, strengthening, or altering feedback loops, by adding new feedback loops. That is why I call these archetypes not just traps, but opportunities.

Policy Resistance—Fixes that Fail

> I think the investment tax credit has a good history of being an effective economic stimulus," said Joseph W. Duncan, chief economist for Dun & Bradstreet Corp. . . .
>
> But skeptics abound. They say nobody can prove any benefit to economic growth from investment credits, which have been granted, altered, and repealed again and again in the last 30 years.
> —John H. Cushman, Jr., *International Herald Tribune*, 1992[2]

As we saw in Chapter Two, the primary symptom of a balancing feedback loop structure is that not much changes, despite outside forces pushing the system. Balancing loops stabilize systems; behavior patterns persist. This is a great structure if you are trying to maintain your body temperature at 37°C (98.6°F), but some behavior patterns that persist over long periods of time are undesirable. Despite efforts to invent technological or policy "fixes," the system seems to be intractably stuck, producing the same behavior every year. This is the systemic trap of "fixes that fail" or "policy resistance." You see this when farm programs try year after year to reduce gluts, but there is still overproduction. There are wars on drugs, after which drugs are as prevalent as ever. There is little evidence that investment tax credits and many other policies designed to stimulate investment when the market is not rewarding investment actually work. No single policy yet has been able to bring down health care costs in the United States. Decades of "job creation" have not managed to keep unemployment permanently low. You probably can name a dozen other areas in which energetic efforts consistently produce non-results.

Policy resistance comes from the bounded rationalities of the actors in a system, each with his or her (or "its" in the case of an institution) own goals. Each actor monitors the state of the system with regard to some important variable—income or prices or housing or drugs or investment—and compares that state with his, her, or its goal. If there is a discrepancy, each actor does something to correct the situation. Usually the greater the discrepancy between the goal and the actual situation, the more emphatic the action will be.

Such resistance to change arises when goals of subsystems are different from and inconsistent with each other. Picture a single-system stock—drug supply on the city streets, for example—with various actors trying to pull that stock in different directions. Addicts want to keep it high, enforcement agencies want to keep it low, pushers want to keep it right in the middle so prices don't get either too high or too low. The average citizen really just wants to be safe from robberies by addicts trying to get money to buy drugs. All the actors work hard to achieve their different goals.

If any one actor gains an advantage and moves the system stock (drug supply) in one direction (enforcement agencies manage to cut drug imports at the border), the others double their efforts to pull it back (street prices go up, addicts have to commit more crimes to buy their daily fixes, higher prices bring more profits, suppliers use the profits to buy planes and boats to evade the border patrols). Together, the countermoves produce a standoff, the stock is not much different from before, and that is not what anybody wants.

In a policy-resistant system with actors pulling in different directions, everyone has to put great effort into keeping the system where no one wants it to be. If any single actor lets up, the others will drag the system closer to their goals, and farther from the goal of the one who let go. In fact, this system structure can operate in a ratchet mode: Intensification of anyone's effort leads to intensification of everyone else's. It's hard to reduce the intensification. It takes a lot of mutual trust to say, OK, why don't we all just back off for a while?

The results of policy resistance can be tragic. In 1967, the Romanian government decided that Romania needed more people and that the way to get them was to make abortions for women under age forty-five illegal. Abortions were abruptly banned. Shortly thereafter, the birth rate tripled. Then the policy resistance of the Romanian people set in.

Although contraceptives and abortions remained illegal, the birth rate slowly came back down nearly to its previous level. This result was achieved primarily though dangerous, illegal abortions, which tripled the maternal mortality rate. In addition, many of the unwanted children that had been born when abortions were illegal were abandoned to orphanages. Romanian families were too poor to raise the many children their government desired decently, and they knew it. So, they resisted the government's pull toward larger family size, at great cost to themselves and to the generation of children who grew up in orphanages.

One way to deal with policy resistance is to try to overpower it. If you wield enough power and can keep wielding it, the power approach can work, at the cost of monumental resentment and the possibility of explosive consequences if the power is ever let up. This is what happened with the formulator of the Romanian population policy, dictator Nicolae Ceausescu, who tried long and hard to overpower the resistance to his policy. When his government was overturned, he was executed, along with his family. The first law the new government repealed was the ban on abortion and contraception.

The alternative to overpowering policy resistance is so counterintuitive that it's usually unthinkable. Let go. Give up ineffective policies. Let the resources and energy spent on both enforcing and resisting be used for more constructive purposes. You won't get your way with the system, but it won't go as far in a bad direction as you think, because much of the action you were trying to correct was in response to your own action. If you calm down, those who are pulling against you will calm down too. This is what happened in 1933 when Prohibition ended in the United States; the alcohol-driven chaos also largely ended.

That calming down may provide the opportunity to look more closely at the feedbacks within the system, to understand the bounded rationality behind them, and to find a way to meet the goals of the participants in the system while moving the state of the system in a better direction.

For example, a nation wanting to increase its birth rate might ask why families are having few children and discover that it isn't because they don't like children. Perhaps they haven't the resources, the living space, the time, or the security to have more. Hungary, at the same time Romania was banning abortions, also was worried about its low birth rate—fearing an economic downturn could result from fewer people in the workforce. The Hungarian government discovered that cramped housing was one reason for small family size. The government devised a policy that rewarded larger families with more living space. This policy was only partially successful, because housing was not the only problem. But it was far more successful than Romania's policy and it avoided Romania's disastrous results.[3]

The most effective way of dealing with policy resistance is to find a way of aligning the various goals of the subsystems, usually by providing an overarching goal that allows all actors to break out of their bounded rationality. If everyone can work harmoniously toward the same outcome (if all feedback loops are serving the same goal), the results can be amazing. The most familiar examples of this harmonization of goals are mobilizations of economies during wartime, or recovery after war or natural disaster.

Another example was Sweden's population policy. During the 1930s, Sweden's birth rate dropped precipitously, and, like the governments of Romania and Hungary, the Swedish government worried about that. Unlike Romania and Hungary, the Swedish government assessed its goals and those of the population and decided that there was a basis of agreement, not on the size of the family, but on the quality of child care. Every child should be wanted and nurtured. No child should be in material need. Every child should have access to excellent education and health care. These were goals around which the government and the people could align themselves.

The resulting policy looked strange during a time of low birth rate, because it included free contraceptives and abortion—because of the principle that every child should be wanted. The policy also included widespread sex education, easier divorce laws, free obstetrical care, support for families in need, and greatly increased investment in education and health care.[4] Since then, the Swedish birth rate has gone up and down several times without causing panic in either direction, because the nation is focused on a far more important goal than the number of Swedes.

Harmonization of goals in a system is not always possible, but it's an option worth looking for. It can be found only by letting go of more narrow goals and considering the long-term welfare of the entire system.

THE TRAP: POLICY RESISTANCE

When various actors try to pull a system stock toward various goals, the result can be policy resistance. Any new policy, especially if it's effective, just pulls the stock farther from the goals of other actors and produces additional resistance, with a result that no one likes, but that everyone expends considerable effort in maintaining.

THE WAY OUT

Let go. Bring in all the actors and use the energy formerly expended on resistance to seek out mutually satisfactory ways for all goals to be realized—or redefinitions of larger and more important goals that everyone can pull toward together.

The Tragedy of the Commons

> Leaders of Chancellor Helmut Kohl's coalition, led by the Christian Democratic Union, agreed last week with the opposition Social Democrats, after months of bickering, to turn back a flood of economic migrants by tightening conditions for claiming asylum.
>
> —*International Herald Tribune*, 1992[5]

The trap called the tragedy of the commons comes about when there is escalation, or just simple growth, in a commonly shared, erodable environment.

Ecologist Garrett Hardin described the commons system in a classic article in 1968. Hardin used as his opening example a common grazing land:

> Picture a pasture open to all. It is to be expected that each herdsman will try to keep as many cattle as possible on the commons.... Explicitly or implicitly, more or less consciously, he asks, "What is the utility to me of adding one more animal to my herd?"...

> Since the herdsman receives all the proceeds from the sale of the additional animal, the positive utility is nearly +1. . . . Since, however, the effects of overgrazing are shared by all, . . . the negative utility for any particular decision-making herdsman is only a fraction of −1. . . .
>
> The rational herdsman concludes that the only sensible course for him to pursue is to add another animal to his herd. And another; and another. . . . But this is the conclusion reached by each and every rational herdsman sharing a commons. Therein is the tragedy. Each . . . is locked into a system that compels him to increase his herd without limit—in a world that is limited. Ruin is the destination toward which all . . . rush, each pursuing his own best interest.[6]

Bounded rationality in a nutshell!

In any commons system there is, first of all, a resource that is commonly shared (the pasture). For the system to be subject to tragedy, the resource must be not only limited, but erodable when overused. That is, beyond some threshold, the less resource there is, the less it is able to regenerate itself, or the more likely it is to be destroyed. As there is less grass on the pasture, the cows eat even the base of the stems from which the new grass grows. The roots no longer hold the soil from washing away in the rains. With less soil, the grass grows more poorly. And so forth. Another reinforcing feedback loop running downhill.

A commons system also needs users of the resource (the cows and their owners), which have good reason to increase, and which increase at a rate *that is not influenced by the condition of the commons.* The individual herdsman has no reason, no incentive, no strong feedback, to let the possibility of overgrazing stop him from adding another cow to the common pasture. To the contrary, he or she has everything to gain.

The hopeful immigrant to Germany expects nothing but benefit from that country's generous asylum laws, and has no reason to take into consideration the fact that too many immigrants will inevitably force Germany to toughen those laws. In fact, the knowledge that Germany is discussing that possibility is all the more reason to hurry to Germany!

The tragedy of the commons arises from *missing (or too long delayed) feedback* from the resource to the growth of the users of that resource.

The more users there are, the more resource is used. The more resource is used, the less there is per user. If the users follow the bounded rationality of the commons ("There's no reason for *me* to be the one to limit my cows!"), there is no reason for any of them to decrease their use. Eventually, then, the harvest rate will exceed the capacity of the resource to bear the harvest. Because there is no feedback to the user, overharvesting will continue. The resource will decline. Finally, the erosion loop will kick in, the resource will be destroyed, and all the users will be ruined.

Surely, you'd think, no group of people would be so shortsighted as to destroy their commons. But consider just a few commonplace examples of commons that are being driven, or have been driven, to disaster:

- Uncontrolled access to a popular national park can bring in such crowds that the park's natural beauties are destroyed.
- It is to everyone's immediate advantage to go on using fossil fuels, although carbon dioxide from these fuels is a greenhouse gas that is causing global climate change.
- If every family can have any number of children it wants, but society as a whole has to support the cost of education, health care, and environmental protection for all children, the number of children born can exceed the capacity of the society to support them all. (This is the example that caused Hardin to write his article.)

These examples have to do with overexploitation of renewable resources—a structure you have seen already in the systems zoo. Tragedy can lurk not only in the use of common resources, but also in the use of common sinks, shared places where pollution can be dumped. A family, company, or nation can reduce its costs, increase its profits, or grow faster if it can get the entire community to absorb or handle its wastes. It reaps a large gain, while itself having to live with only a fraction of its own pollution (or none, if it can dump downstream or downwind). There is no rational reason why a polluter should desist. In these cases, the feedback influencing the rate of use of the common resource—whether it is a source or a sink—is weak.

If you think that the reasoning of an exploiter of the commons is hard to understand, ask yourself how willing you are to carpool in order to reduce air pollution, or to clean up after yourself whenever you make a mess.

The structure of a commons system makes selfish behavior much more convenient and profitable than behavior that is responsible to the whole community and to the future.

There are three ways to avoid the tragedy of the commons.

- *Educate and exhort.* Help people to see the consequences of unrestrained use of the commons. Appeal to their morality. Persuade them to be temperate. Threaten transgressors with social disapproval or eternal hellfire.
- *Privatize the commons.* Divide it up, so that each person reaps the consequences of his or her own actions. If some people lack the self-control to stay below the carrying capacity of their own private resource, those people will harm only themselves and not others.
- *Regulate the commons.* Garrett Hardin calls this option, bluntly, "mutual coercion, mutually agreed upon." Regulation can take many forms, from outright bans on certain behaviors to quotas, permits, taxes, incentives. To be effective, regulation must be enforced by policing and penalties.

The first of these solutions, exhortation, tries to keep use of the commons low enough through moral pressure that the resource is not threatened. The second, privatization, makes a direct feedback link from the condition of the resource to those who use it, by making sure that gains and losses fall on the same decision maker. The owner still may abuse the resource, but now it takes ignorance or irrationality to do so. The third solution, regulation, makes an indirect feedback link from the condition of the resource through regulators to users. For this feedback to work, the regulators must have the expertise to monitor and interpret correctly the condition of the commons, they must have effective means of deterrence, and they must have the good of the whole community at heart. (They cannot be uninformed or weak or corrupt.)

Some "primitive" cultures have managed common resources effectively for generations through education and exhortation. Garrett Hardin does not believe that option is dependable, however. Common resources protected only by tradition or an "honor system" may attract those who do not respect the tradition or who have no honor.

Privatization works more reliably than exhortation, if society is willing to let some individuals learn the hard way. But many resources, such as the atmosphere and the fish of the sea, simply cannot be privatized. That leaves only the option of "mutual coercion, mutually agreed upon."

Life is full of mutual-coercion arrangements, most of them so ordinary you hardly stop to think about them. Every one of them limits the freedom to abuse a commons, while preserving the freedom to use it. For example:

- The common space in the center of a busy intersection is regulated by traffic lights. You can't drive through whenever you want to. When it is your turn, however, you can pass through more safely than would be possible if there were an unregulated free-for-all.
- Use of common parking spaces in downtown areas are parceled out by meters, which charge for a space and limit the time it can be occupied. You are not free to park wherever you want for as long as you want, but you have a higher chance of finding a parking space than you would if the meters weren't there.
- You may not help yourself to the money in a bank, however advantageous it might be for you to do so. Protective devices such as strongboxes and safes, reinforced by police and jails, prevent you from treating a bank as a commons. In return, your own money in the bank is protected.
- You may not broadcast at will over the wavelengths that carry radio or television signals. You must obtain a permit from a regulatory agency. If your freedom to broadcast were not limited, the airwaves would be a chaos of overlapping signals.
- Many municipal garbage systems have become so expensive that households are now charged for garbage disposal depending on the amount of garbage they generate—transforming the previous commons to a regulated pay-as-you-go system.

Notice from these examples how many different forms "mutual coercion, mutually agreed upon" can take. The traffic light doles out access to the commons on a "take your turn" basis. The meters charge for use of the parking commons. The bank uses physical barriers and strong penalties. Permits to use broadcasting frequencies are issued to applicants by a government agency. And garbage fees directly restore the missing feedback, letting each household feel the economic impact of its own use of the commons.

Most people comply with regulatory systems most of the time, as long as they are mutually agreed upon and their purpose is understood. But all regulatory systems must use police power and penalties for the occasional noncooperator.

THE TRAP: TRAGEDY OF THE COMMONS

When there is a commonly shared resource, every user benefits directly from its use, but shares the costs of its abuse with everyone else. Therefore, there is very weak feedback from the condition of the resource to the decisions of the resource users. The consequence is overuse of the resource, eroding it until it becomes unavailable to anyone.

THE WAY OUT

Educate and exhort the users, so they understand the consequences of abusing the resource. And also restore or strengthen the missing feedback link, either by privatizing the resource so each user feels the direct consequences of its abuse or (since many resources cannot be privatized) by regulating the access of all users to the resource.

Drift to Low Performance

In this recession, the British have discovered that . . . the economy is just as downwardly mobile as ever. Even national disasters are now seized on as portents of further decline. The *Independent* on Sunday carried a front-page article on "the ominous feeling that the Windsor fire is symptomatic of the country at large, that it stems from the new national characteristic of ineptitude. . . ."

Insisted Lord Peston, Labor's trade and industry spokesman, "We know what we ought to do, for some reason we just don't do it."

> Politicians, businessmen, and economists fault the country as a place where the young receive substandard education, where both labor and management are underskilled, where investment is skimped, and where politicians mismanage the economy.
> —Erik Ipsen, *International Herald Tribune*, 1992 [7]

Some systems not only resist policy and stay in a normal bad state, they keep getting worse. One name for this archetype is "drift to low performance." Examples include falling market share in a business, eroding quality of service at a hospital, continuously dirtier rivers or air, increased fat in spite of periodic diets, the state of America's public schools—or my one-time jogging program, which somehow just faded away.

The actor in this feedback loop (British government, business, hospital, fat person, school administrator, jogger) has, as usual, a performance goal or desired system state that is compared to the actual state. If there is a discrepancy, action is taken. So far, that is an ordinary balancing feedback loop that should keep performance at the desired level.

But in this system, there is a distinction between the actual system state and the perceived state. *The actor tends to believe bad news more than good news.* As actual performance varies, the best results are dismissed as aberrations, the worst results stay in the memory. The actor thinks things are worse than they really are.

And to complete this tragic archetype, *the desired state of the system is influenced by the perceived state.* Standards aren't absolute. When perceived performance slips, the goal is allowed to slip. "Well, that's about all you can expect." "Well, we're not doing much worse than we were last year." "Well, look around, everybody else is having trouble too."

The balancing feedback loop that should keep the system state at an acceptable level is overwhelmed by a reinforcing feedback loop heading downhill. The lower the perceived system state, the lower the desired state. The lower the desired state, the less discrepancy, and the less corrective action is taken. The less corrective action, the lower the system state. If this loop is allowed to run unchecked, it can lead to a continuous degradation in the system's performance.

Another name for this system trap is "eroding goals." It is also called the "boiled frog syndrome," from the old story (I don't know whether it is true) that a frog put suddenly in hot water will jump right out, but if it is put into cold water that is gradually heated up, the frog will stay there happily until it boils. "Seems to be getting a little warm in here. Well, but then it's not so much warmer than it was a while ago." Drift to low performance is a gradual process. If the system state plunged quickly, there would be an agitated corrective process. But if it drifts down slowly enough to erase the memory of (or belief in) how much better things used to be, everyone is lulled into lower and lower expectations, lower effort, lower performance.

There are two antidotes to eroding goals. One is to keep standards absolute, regardless of performance. Another is to make goals sensitive to the *best* performances of the past, instead of the worst. If perceived performance has an upbeat bias instead of a downbeat one, if one takes the best results as a standard, and the worst results only as a temporary setback, then the same system structure can pull the system up to better and better performance. The reinforcing loop going downward, which said "the worse things get, the worse I'm going to let them get," becomes a reinforcing loop going upward: "The better things get, the harder I'm going to work to make them even better."

If I had applied that lesson to my jogging, I'd be running marathons by now.

THE TRAP: DRIFT TO LOW PERFORMANCE

Allowing performance standards to be influenced by past performance, especially if there is a negative bias in perceiving past performance, sets up a reinforcing feedback loop of eroding goals that sets a system drifting toward low performance.

THE WAY OUT

Keep performance standards absolute. Even better, let standards be enhanced by the best actual performances instead of being discouraged by the worst. Use the same structure to set up a drift toward high performance!

Escalation

> Islamic militants kidnapped an Israeli soldier Sunday and threat-
> ened to kill him unless the army quickly releases the imprisoned
> founder of a dominant Muslim group in the Gaza Strip. . . . The
> kidnapping . . . came in a wave of intense violence, . . . with the
> shooting of three Palestinians and an Israeli soldier who . . . was
> gunned down from a passing vehicle while he was on patrol in a
> jeep. In addition Gaza was buffeted by repeated clashes between
> stone-throwing demonstrators and Israeli troops, who opened fire
> with live ammunition and rubber bullets, wounding at least 120
> people.
>
> —Clyde Haberman, *International Herald Tribune*, 1992[8]

I already mentioned one example of escalation early in this book; the system of kids fighting. You hit me, so I hit you back a little harder, so you hit me back a little harder, and pretty soon we have a real fight going.

"I'll raise you one" is the decision rule that leads to escalation. Escalation comes from a reinforcing loop set up by competing actors trying to get ahead of each other. The goal of one part of the system or one actor is not absolute, like the temperature of a room thermostat being set at 18°C (65°F), but is related to the state of another part of the system, another actor. Like many of the other system traps, escalation is not necessarily a bad thing. If the competition is about some desirable goal, like a more effi-cient computer or a cure for AIDS, it can hasten the whole system toward the goal. But when it is escalating hostility, weaponry, noise, or irritation, this is an insidious trap indeed. The most common and awful examples are arms races and those places on earth where implacable enemies live constantly on the edge of self-reinforcing violence.

Each actor takes its desired state from the other's perceived system state—and ups it! Escalation is not just keeping up with the Joneses, but keeping slightly ahead of the Joneses. The United States and the Soviet Union for years exaggerated their reports of each other's armaments in order to justify more armaments of their own. Each weapons increase on one side caused a scramble to surpass it on the other side. Although each side blamed the other for the escalation, it would be more systematic to say that each side was escalating itself—its own weapons development started a process that was sure to require still more weapons development in the future. This system caused trillions of dollars of expense, the degradation of the economies of two superpowers, and the evolution of unimaginably destructive weapons, which still threaten the world.

Negative campaigning is another perverse example of escalation. One candidate smears another, so the other smears back, and so forth, until the voters have no idea that their candidates have any positive features, and the whole democratic process is demeaned.

Then there are price wars, with one economic competitor underpricing another, which causes the other to cut prices more, which causes the first to cut prices yet again, until both sides are losing money, but neither side can easily back out. This kind of escalation can end with the bankruptcy of one of the competitors.

Advertising companies escalate their bids for the attention of the consumer. One company does something bright and loud and arresting. Its competitor does something louder, bigger, brasher. The first company outdoes that. Advertising becomes ever more present in the environment (in the mail, on the telephone), more garish, more noisy, more intrusive, until the consumer's senses are dulled to the point at which almost no advertiser's message can penetrate.

The escalation system also produces the increasing loudness of conversation at cocktail parties, the increasing length of limousines, and the increasing raunchiness of rock bands.

Escalation also could be about peacefulness, civility, efficiency, subtlety, quality. But even escalating in a good direction can be a problem, because it isn't easy to stop. Each hospital trying to outdo the others in up-to-date, powerful, expensive diagnostic machines can lead to out-of-sight health care costs. Escalation in morality can lead to holier-than-thou sanctimoniousness. Escalation in art can lead from baroque to rococo to kitsch. Escalation in environmentally responsible lifestyles can lead to rigid and unnecessary puritanism.

Escalation, being a reinforcing feedback loop, builds exponentially. Therefore, it can carry a competition to extremes faster than anyone would believe possible. If nothing is done to break the loop, the process usually ends with one or both of the competitors breaking down.

One way out of the escalation trap is unilateral disarmament—deliberately reducing your own system state to induce reductions in your competitor's state. Within the logic of the system, this option is almost unthinkable. But it actually can work, if one does it with determination, and if one can survive the short-term advantage of the competitor.

The only other graceful way out of the escalation system is to negotiate a disarmament. That's a structural change, an exercise in system design. It creates a new set of balancing controlling loops to keep the competition in bounds (parental pressure to stop the kids' fight; regulations on the size and placement of advertisements; peace-keeping troops in violence-prone areas). Disarmament agreements in escalation systems are not usually easy to get, and are never very pleasing to the parties involved, but they are much better than staying in the race.

THE TRAP: ESCALATION

When the state of one stock is determined by trying to surpass the state of another stock—and vice versa—then there is a reinforcing feedback loop carrying the system into an arms race, a wealth race, a smear campaign, escalating loudness, escalating violence. The escalation is exponential and can lead to extremes surprisingly quickly. If nothing is done, the spiral will be stopped by someone's collapse—because exponential growth cannot go on forever.

THE WAY OUT

The best way out of this trap is to avoid getting in it. If caught in an escalating system, one can refuse to compete (unilaterally disarm), thereby interrupting the reinforcing loop. Or one can negotiate a new system with balancing loops to control the escalation.

Success to the Successful—Competitive Exclusion

Extremely rich people—the top slice of the top 1 percent of taxpayers—have considerable flexibility to expose less of their income to taxation. . . . Those who can have raced to take bonuses now rather than next year [when taxes are expected to be higher], to cash in stock options, . . . and to move income forward in any way possible.

—Sylvia Nasar, *International Herald Tribune*, 1992[9]

Using accumulated wealth, privilege, special access, or inside information to create more wealth, privilege, access or information are examples of the archetype called "success to the successful." This system trap is found whenever the winners of a competition receive, as part of the reward, the means to compete even more effectively in the future. That's a reinforcing feedback loop, which rapidly divides a system into winners who go on winning, and losers who go on losing.

Anyone who has played the game of Monopoly knows the success-to-the-successful system. All players start out equal. The ones who manage to be first at building "hotels" on their property are able to extract "rent" from the other players—which they can then use to buy more hotels. The more hotels you have, the more hotels you can get. The game ends when one player has bought up everything, unless the other players have long ago quit in frustration.

Once our neighborhood had a contest with a $100 reward for the family that put up the most impressive display of outdoor Christmas lights. The family that won the first year spent the $100 on more Christmas lights. After that family won three years in a row, with their display getting more elaborate every year, the contest was suspended.

To him that hath shall be given. The more the winner wins, the more he, she, or it can win in the future. If the winning takes place in a limited environment, such that everything the winner wins is extracted from the losers, the losers are gradually bankrupted, or forced out, or starved.

Success to the successful is a well-known concept in the field of ecology, where it is called "the competitive exclusion principle." This principle says that two different species cannot live in exactly the same ecological niche, competing for exactly the same resources. Because the two species are different, one will necessarily reproduce faster, or be able to use the resource more efficiently than the other. It will win a larger share of the resource, which will give it the ability to multiply more and keep winning. It will not only dominate the niche, it will drive the losing competitor to extinction. That will happen not by direct confrontation usually, but by appropriating all the resource, leaving none for the weaker competitor.

Another expression of this trap was part of the critique of capitalism by Karl Marx. Two firms competing in the same market will exhibit the same behavior as two species competing in a niche. One will gain a slight advantage, through greater efficiency or smarter investment or better technology or bigger bribes, or whatever. With that advantage, the firm will have more income to invest in productive facilities or newer technologies or advertising or bribes. Its reinforcing feedback loop of capital accumulation will be able to turn faster than that of the other firm, enabling it to produce still more and earn still more. If there is a finite market and no antitrust law to stop it, one firm will take over everything as long as it chooses to reinvest in and expand its production facilities.

Some people think the fall of the communist Soviet Union has disproved the theories of Karl Marx, but this particular analysis of his—that market competition systematically eliminates market competition—is demonstrated wherever there is, or used to be, a competitive market. Because of the reinforcing feedback loop of success to the successful, the many automobile companies in the United States were reduced to three (not one, because of antitrust laws). In most major U.S. cities, there is only one newspaper left. In every market economy, we see long-term trends of declining numbers of farms, while the size of farms increases.

The trap of success to the successful does its greatest damage in the many ways it works to make the rich richer and the poor poorer. Not only do the rich have more ways to avoid taxation than the poor, but:

- In most societies, the poorest children receive the worst educations in the worst schools, if they are able to go to school at all. With few marketable skills, they qualify only for low-paying jobs, perpetuating their poverty.[10]
- People with low income and few assets are not able to borrow from most banks. Therefore, either they can't invest in capital improvements, or they must go to local money-lenders who charge exorbitant interest rates. Even when interest rates are reasonable, the poor pay them, the rich collect them.
- Land is held so unevenly in many parts of the world that most farmers are tenants on someone else's land. They must pay part of their crops to the landowner for the privilege of work-

ing the land, and so never are able to buy land of their own. The landowner uses the income from tenants to buy more land.

Those are only a few of the feedbacks that perpetuate inequitable distribution of income, assets, education, and opportunity. Because the poor can afford to buy only small quantities (of food, fuel, seed, fertilizer), they pay the highest prices. Because they are often unorganized and inarticulate, a disproportionately small part of government expenditure is allocated to their needs. Ideas and technologies come to them last. Disease and pollution come to them first. They are the people who have no choice but to take dangerous, low-paying jobs, whose children are not vaccinated, who live in crowded, crime-prone, disaster-prone areas.

How do you break out of the trap of success to the successful?

Species and companies sometimes escape competitive exclusion by *diversifying*. A species can learn or evolve to exploit new resources. A company can create a new product or service that does not directly compete with existing ones. Markets tend toward monopoly and ecological niches toward monotony, but they also create offshoots of diversity, new markets, new species, which in the course of time may attract competitors, which then begin to move the system toward competitive exclusion again.

Diversification is not guaranteed, however, especially if the monopolizing firm (or species) has the power to crush all offshoots, or buy them up, or deprive them of the resources they need to stay alive. Diversification doesn't work as a strategy for the poor.

The success-to-the-successful loop can be kept under control by putting into place feedback loops that keep any competitor from taking over entirely. That's what antitrust laws do in theory and sometimes in practice. (One of the resources very big companies can win by winning, however, is the power to weaken the administration of antitrust laws.)

The most obvious way out of the success-to-the-successful archetype is by periodically "leveling the playing field." Traditional societies and game designers instinctively design into their systems some way of equalizing advantages, so the game stays fair and interesting. Monopoly games start over again with everyone equal, so those who lost last time have a chance to win. Many sports provide handicaps for weaker players. Many traditional societies have some version of the Native American "potlatch," a ritual in which those who have the most give away many of their possessions to those who have the least.

There are many devices to break the loop of the rich getting richer and the poor getting poorer: tax laws written (unbeatably) to tax the rich at higher rates than the poor; charity; public welfare; labor unions; universal and equal health care and education; taxation on inheritance (a way of starting the game over with each new generation). Most industrial societies have some combination of checks like these on the workings of the success-to-the-successful trap, in order to keep everyone in the game. Gift-giving cultures redistribute wealth through potlatches and other ceremonies that increase the social standing of the gift giver.

These equalizing mechanisms may derive from simple morality, or they may come from the practical understanding that losers, if they are unable to get out of the game of success to the successful, and if they have no hope of winning, could get frustrated enough to destroy the playing field.

THE TRAP: SUCCESS TO THE SUCCESSFUL

If the winners of a competition are systematically rewarded with the means to win again, a reinforcing feedback loop is created by which, if it is allowed to proceed uninhibited, the winners eventually take all, while the losers are eliminated.

THE WAY OUT

Diversification, which allows those who are losing the competition to get out of that game and start another one; strict limitation on the fraction of the pie any one winner may win (antitrust laws); policies that level the playing field, removing some of the advantage of the strongest players or increasing the advantage of the weakest; policies that devise rewards for success that do not bias the next round of competition.

Shifting the Burden to the Intervenor—Addiction

You get some sense of what an incredible downward spiral we're in. Because more costs keep being shifted to the private sector, more private sector people stop insuring their employees. We are . . . now up to 100,000 Americans a month losing their health insurance.

An enormous percentage of them qualify for state Medicaid benefits. And since states can't run a deficit, they all go out and either underfund education, or underfund children's investment programs, or raise taxes, and that takes money away from other investments.

—Bill Clinton, *International Herald Tribune*, 1992[11]

If you want to make a Somali angry, it is said, take away his khat.. . .

Khat is the fresh tender leaves and twigs of the *catha edulis* plant. . . . It is pharmacologically related to amphetamines. . . .

Abdukadr Mahmoud Farah, 22, said he first started chewing khat when he was 15. . . . "The reason is not to think of this place. When I use it, I get happy. I can do everything. I do not get tired."

—Keith B. Richburg, *International Herald Tribune*, 1992[12]

Most people understand the addictive properties of alcohol, nicotine, caffeine, sugar, and heroin. Not everyone recognizes that addiction can appear in larger systems and in other guises—such as the dependence of industry on government subsidy, the reliance of farmers on fertilizers, the addiction of Western economies to cheap oil or weapons manufacturers to government contracts.

This trap is known by many names: addiction, dependence; shifting the burden to the intervenor. The structure includes a stock with in-flows and out-flows. The stock can be physical (a crop of corn) or *meta*-physical (a sense of well-being or self-worth). The stock is maintained by an actor adjusting a balancing feedback loop—either altering the in-flows or out-flows. The actor has a goal and compares it with a perception of the actual state of the stock to determine what action to take.

Say you are a young boy, living in a land of famine and war, and your goal is to boost your sense of well-being so you feel happy and energetic and fearless. There is a huge discrepancy between your desired and actual state, and there are very few options available to you for closing that gap. But one thing you can do is take drugs. The drugs do nothing to improve your real situation—in fact, they likely make it worse. But the drugs quickly alter your *perception* of your state, numbing your senses and making you feel tireless and brave.

Similarly, if you are running an ineffective company, and if you can get the government to subsidize you, you can go on making money and continue to have a good profit, thereby remaining a respected member of society. Or perhaps you are a farmer trying to increase your corn crop on overworked land. You apply fertilizers and get a bumper crop without doing anything to improve the fertility of the soil.

The trouble is that the states created by interventions don't last. The intoxication wears off. The subsidy is spent. The fertilizer is consumed or washed away.

Examples of dependence and burden-shifting systems abound:

- Care of the aged used to be carried on by families, not always easily. So along came Social Security, retirement communities, nursing homes. Now most families no longer have the space, the time, the skills, or the willingness to care for their elderly members.
- Long-distance shipping was carried by railroads and short-distance commuting by subways and streetcars, until the government decided to help out by building highways.
- Kids used to be able to do arithmetic in their heads or with paper and pencil, before the widespread use of calculators.
- Populations built up a partial immunity to diseases such as smallpox, tuberculosis, and malaria, until vaccinations and drugs came along.
- Modern medicine in general has shifted the responsibility for health away from the practices and lifestyle of each individual and onto intervening doctors and medicines.

Shifting a burden to an intervenor can be a good thing. It often is done purposefully, and the result can be an increased ability to keep the system in a desirable state. Surely the 100 percent protection from smallpox vaccines, if it lasts, is preferable to only partial protection from natural smallpox immunity. Some systems really need an intervenor!

But the intervention can become a system trap. A corrective feedback process within the system is doing a poor (or even so-so) job of maintaining the state of the system. A well-meaning and efficient intervenor watches the struggle and steps in to take some of the load. The intervenor quickly brings the system to the state everybody wants it to be in. Congratulations are in order, usually self-congratulations by the intervenor to the intervenor.

Then the original problem reappears, since nothing has been done to solve it at its root cause. So the intervenor applies more of the "solution," disguising the real state of the system again, and thereby failing to act on the problem. That makes it necessary to use still more "solution."

The trap is formed if the intervention, whether by active destruction or simple neglect, undermines the original capacity of the system to maintain itself. If that capability atrophies, then more of the intervention is needed to achieve the desired effect. That weakens the capability of the original system still more. The intervenor picks up the slack. And so forth.

Why does anyone enter the trap? First, the intervenor may not foresee that the initial urge to help out a bit can start a chain of events that leads to ever-increasing dependency, which ultimately will strain the capacity of the intervenor. The American health-care system is experiencing the strains of that sequence of events.

Second, the individual or community that is being helped may not think through the long-term loss of control and the increased vulnerability that go along with the opportunity to shift a burden to an able and powerful intervenor.

If the intervention is a drug, you become addicted. The more you are sucked into an addictive action, the more you are sucked into it again. One definition of addiction used in Alcoholics Anonymous is repeating the same stupid behavior over and over and over, and somehow expecting different results.

Addiction is finding a quick and dirty solution to the *symptom* of the problem, which prevents or distracts one from the harder and longer-term task of solving the real problem. Addictive policies are insidious, because they are so easy to sell, so simple to fall for.

Are insects threatening the crops? Rather than examine the farming methods, the monocultures, the destruction of natural ecosystem controls that have led to the pest outbreak, just apply pesticides. That will make the bugs go away, and allow more monocultures, more destruction of ecosystems. That will bring back the bugs in greater outbursts, requiring more pesticides in the future.

Is the price of oil going up? Rather than acknowledge the inevitable depletion of a nonrenewable resource and increase fuel efficiency or switch to other fuels, we can *fix the price*. (Both the Soviet Union and the United States did this as their first response to the oil-price shocks of the 1970s.) That way we can pretend that nothing is happening and go on burning oil—making the depletion problem worse. When that policy breaks down, we can go to war for oil. Or find more oil. Like a drunk ransacking the house in hopes of unearthing just one more bottle, we can pollute the beaches and invade the last wilderness areas, searching for just one more big deposit of oil.

Breaking an addiction is painful. It may be the physical pain of heroin withdrawal, or the economic pain of a price increase to reduce oil consumption, or the consequences of a pest invasion while natural predator populations are restoring themselves. Withdrawal means finally confronting the real (and usually much deteriorated) state of the system and taking the actions that the addiction allowed one to put off. Sometimes the withdrawal can be done gradually. Sometimes a nonaddictive policy can be put in place first to restore the degraded system with a minimum of turbulence (group support to restore the self-image of the addict, home insulation and high-mileage cars to reduce oil expense, polyculture and crop rotation to reduce crop vulnerability to pests). Sometimes there's no way out but to go cold turkey and just bear the pain.

It's worth going through the withdrawal to get back to an unaddicted state, but it is far preferable to avoid addiction in the first place.

The problem can be avoided up front by intervening in such a way as *to strengthen the ability of the system to shoulder its own burdens*. This option, helping the system to help itself, can be much cheaper and easier than taking over and running the system—something liberal politicians don't seem to understand. The secret is to begin not with a heroic takeover, but with a series of questions.

- Why are the natural correction mechanisms failing?
- How can obstacles to their success be removed?
- How can mechanisms for their success be made more effective?

THE TRAP: SHIFTING THE BURDEN TO THE INTERVENOR

Shifting the burden, dependence, and addiction arise when a solution to a systemic problem reduces (or disguises) the symptoms, but does nothing to solve the underlying problem. Whether it is a substance that dulls one's perception or a policy that hides the underlying trouble, the drug of choice interferes with the actions that could solve the real problem.

If the intervention designed to correct the problem causes the self-maintaining capacity of the original system to atrophy or erode, then a destructive reinforcing feedback loop is set in motion. The system deteriorates; more and more of the solution is then required. The system will become more and more dependent on the intervention and less and less able to maintain its own desired state.

THE WAY OUT

Again, the best way out of this trap is to avoid getting in. Beware of symptom-relieving or signal-denying policies or practices that don't really address the problem. Take the focus off short-term relief and put it on long-term restructuring.

If you are the intervenor, work in such a way as to restore or enhance the system's own ability to solve its problems, then remove yourself.

If you are the one with an unsupportable dependency, build your system's own capabilities back up before removing the intervention. Do it right away. The longer you wait, the harder the withdrawal process will be.

Rule Beating

CALVIN: OK, Hobbes, I've got a plan.

HOBBES: Yeah?

CALVIN: If I do ten spontaneous acts of good will a day from now until Christmas, Santa will have to be lenient in judging the rest of this last year. I can claim I've turned a new leaf.

HOBBES: Well, here's your chance. Susie's coming this way.

CALVIN: Maybe I'll start tomorrow and do 20 a day.

—*International Herald Tribune*, 1992[13]

Wherever there are rules, there is likely to be rule beating. Rule beating means evasive action to get around the intent of a system's rules—abiding by the letter, but not the spirit, of the law. Rule beating becomes a problem only when it leads a system into large distortions, unnatural behaviors that would make no sense at all in the absence of the rules. If it gets out of hand, rule beating can cause systems to produce very damaging behavior indeed.

Rule beating that distorts nature, the economy, organizations, and the human spirit can be destructive. Here are some examples, some serious, some less so, of rule beating:

- Departments of governments, universities, and corporations often engage in pointless spending at the end of the fiscal year just to get rid of money—because if they don't spend their budget this year, they will be allocated less next year.
- In the 1970s, the state of Vermont adopted a land-use law called Act 250 that requires a complex approval process for subdivisions that create lots of ten acres or less. Now Vermont has an extraordinary number of lots just a little over ten acres.
- To reduce grain imports and assist local grain farmers, European countries imposed import restrictions on feed grains in the 1960s. No one thought, while the restrictions were being drafted, about the starchy root called cassava, which also happens to be a good animal feed. Cassava was not included in the restrictions. So corn imports from North America were replaced by cassava imports from Asia.[14]

• The U.S. Endangered Species Act restricts development wherever an endangered species has its habitat. Some landowners, on discovering that their property harbors an endangered species, purposely hunt or poison it, so the land can be developed.

Notice that rule beating produces the *appearance* of rules being followed. Drivers obey the speed limits, when they're in the vicinity of a police car. Feed grains are no longer imported into Europe. Development does not proceed where an endangered species is documented as present. The "letter of the law" is met, the spirit of the law is not. That is a warning about needing to design the law with the whole system, including its self-organizing evasive possibilities, in mind.

Rule beating is usually a response of the lower levels in a hierarchy to overrigid, deleterious, unworkable, or ill-defined rules from above. There are two generic responses to rule beating. One is to try to stamp out the self-organizing response by strengthening the rules or their enforcement—usually giving rise to still greater system distortion. That's the way further into the trap.

The way out of the trap, the opportunity, is to understand rule beating as useful feedback, and to revise, improve, rescind, or better explain the rules. Designing rules better means foreseeing as far as possible the effects of the rules on the subsystems, including any rule beating they might engage in, and structuring the rules to turn the self-organizing capabilities of the system in a positive direction.

THE TRAP: RULE BEATING

Rules to govern a system can lead to rule beating—perverse behavior that gives the appearance of obeying the rules or achieving the goals, but that actually distorts the system.

THE WAY OUT

Design, or redesign, rules to release creativity not in the direction of beating the rules, but in the direction of achieving the purpose of the rules.

Seeking the Wrong Goal

The government formally acknowledged Friday what private economists have been saying for months: Japan will not come close to hitting the 3.5 percent growth target government planners set a year ago. . . .

GNP grew in 1991 by 3.5 percent and in 1990 by 5.5 percent. Since the beginning of this fiscal year . . . the economy has been stagnant or contracting. . . .

Now that the forecast . . . has been lowered sharply, pressure from politicians and business is likely to grow on the Finance Ministry to take stimulative measures.

—*International Herald Tribune*, 1992[15]

Back in Chapter One, I said that one of the most powerful ways to influence the behavior of a system is through its purpose or goal. That's because the goal is the direction-setter of the system, the definer of discrepancies that require action, the indicator of compliance, failure, or success toward which balancing feedback loops work. If the goal is defined badly, if it doesn't measure what it's supposed to measure, if it doesn't reflect the real welfare of the system, then the system can't possibly produce a desirable result. Systems, like the three wishes in the traditional fairy tale, have a terrible tendency to produce exactly and only what you ask them to produce. Be careful what you ask them to produce.

If the desired system state is national security, and that is defined as the amount of money spent on the military, the system will produce military spending. It may or may not produce national security. In fact, security may be undermined if the spending drains investment from other parts of the economy, and if the spending goes for exorbitant, unnecessary, or unworkable weapons.

If the desired system state is good education, measuring that goal by the amount of money spent per student will ensure money spent per student. If the quality of education is measured by performance on standardized tests, the system will produce performance on standardized tests. Whether either of these measures is correlated with good education is at least worth thinking about.

In the early days of family planning in India, program goals were defined in terms of the number of IUDs implanted. So doctors, in their eagerness to meet their targets, put loops into women without patient approval.

These examples confuse effort with result, one of the most common mistakes in designing systems around the wrong goal. Maybe the worst mistake of this kind has been the adoption of the GNP as the measure of national economic success. The GNP is the gross national product, the money value of the final goods and services produced by the economy. As a measure of human welfare, it has been criticized almost from the moment it was invented:

> The gross national product does not allow for the health of our children, the quality of their education or the joy of their play. It does not include the beauty of our poetry or the strength of our marriages, the intelligence of our public debate or the integrity of our public officials. It measures neither our wit nor our courage, neither our wisdom nor our learning, neither our compassion nor our devotion to our country, it measures everything in short, except that which makes life worthwhile.[16]

> We have a system of national accounting that bears no resemblance to the national economy whatsoever, for it is not the record of our life at home but the fever chart of our consumption.[17]

The GNP lumps together goods and bads. (If there are more car accidents and medical bills and repair bills, the GNP goes up.) It counts only marketed goods and services. (If all parents hired people to bring up their children, the GNP would go up.) It does not reflect distributional equity. (An expensive second home for a rich family makes the GNP go up more than an inexpensive basic home for a poor family.) It measures effort rather than achievement, gross production and consumption rather than efficiency. New light bulbs that give the same light with one-eighth the electricity and that last ten times as long make the GNP go down.

GNP is a measure of *throughput*—flows of stuff made and purchased in a year—rather than capital stocks, the houses and cars and computers and stereos that are the source of real wealth and real pleasure. It could be argued that the best society would be one in which capital stocks can be maintained and used with the lowest possible throughput, rather than the highest.

Although there is every reason to want a thriving economy, there is no particular reason to want the GNP to go up. But governments around the world respond to a signal of faltering GNP by taking numerous actions to keep it growing. Many of those actions are simply wasteful, stimulating inefficient production of things no one particularly wants. Some of them, such as overharvesting forests in order to stimulate the economy in the short term, threaten the long-term good of the economy or the society or the environment.

If you define the goal of a society as GNP, that society will do its best to produce GNP. It will not produce welfare, equity, justice, or efficiency unless you define a goal and regularly measure and report the state of welfare, equity, justice, or efficiency. The world would be a different place if instead of competing to have the highest per capita GNP, nations competed to have the highest per capita stocks of wealth with the lowest throughput, or the lowest infant mortality, or the greatest political freedom, or the cleanest environment, or the smallest gap between the rich and the poor.

Seeking the wrong goal, satisfying the wrong indicator, is a system characteristic almost opposite from rule beating. In rule beating, the system is out to evade an unpopular or badly designed rule, while giving the appearance of obeying it. In seeking the wrong goal, the system obediently follows the rule and produces its specified result—which is not necessarily what anyone actually wants. You have the problem of wrong goals when you find

THE TRAP: SEEKING THE WRONG GOAL

System behavior is particularly sensitive to the goals of feedback loops. If the goals—the indicators of satisfaction of the rules—are defined inaccurately or incompletely, the system may obediently work to produce a result that is not really intended or wanted.

THE WAY OUT

Specify indicators and goals that reflect the real welfare of the system. Be especially careful not to confuse effort with result or you will end up with a system that is producing effort, not result.

something stupid happening "because it's the rule." You have the problem of rule beating when you find something stupid happening because it's the way around the rule. Both of these system perversions can be going on at the same time with regard to the same rule.

INTERLUDE • *The Goal of Sailboat Design*

Once upon a time, people raced sailboats not for millions of dollars or for national glory, but just for the fun of it.

They raced the boats they already had for normal purposes, boats that were designed for fishing, or transporting goods, or sailing around on weekends.

It quickly was observed that races are more interesting if the competitors are roughly equal in speed and maneuverability. So rules evolved, that defined various classes of boat by length and sail area and other parameters, and that restricted races to competitors of the same class.

Soon boats were being designed not for normal sailing, but for winning races within the categories defined by the rules. They squeezed the last possible burst of speed out of a square inch of sail, or the lightest possible load out of a standard-sized rudder. These boats were strange-looking and strange-handling, not at all the sort of boat you would want to take out fishing or for a Sunday sail. As the races became more serious, the rules became stricter and the boat designs more bizarre.

Now racing sailboats are extremely fast, highly responsive, and nearly unseaworthy. They need athletic and expert crews to manage them. No one would think of using an America's Cup yacht for any purpose other than racing within the rules. The boats are so optimized around the present rules that they have lost all resilience. Any change in the rules would render them useless.

PART THREE
Creating Change—in Systems and in Our Philosophy

SIX

Leverage Points—
Places to Intervene in a System

IBM . . . announced 25,000 new job cuts and a large reduction in
spending on research. . . . Spending on development research is to
be lowered by $1 billion next year. . . . Chairman John K. Akers . . .
said IBM was still a world and industry leader in research but felt it
could do better by "shifting to areas for growth," meaning services,
which need less capital but also return less profit in the long run.

—Lawrence Malkin, *International Herald Tribune*, 1992[1]

So, how do we change the structure of systems to produce more of what
we want and less of that which is undesirable? After years of working with
corporations on their systems problems, MIT's Jay Forrester likes to say
that the average manager can define the current problem very cogently,
identify the system structure that leads to the problem, and guess with great
accuracy where to look for leverage points—places in the system where a
small change could lead to a large shift in behavior.

This idea of leverage points is not unique to systems analysis—it's
embedded in legend: the silver bullet; the trimtab; the miracle cure; the
secret passage; the magic password; the single hero who turns the tide of
history; the nearly effortless way to cut through or leap over huge obstacles.
We not only want to believe that there are leverage points, we want to know
where they are and how to get our hands on them. Leverage points are
points of power.

But Forrester goes on to point out that although people deeply involved
in a system often know intuitively where to find leverage points, more often
than not they push the change in the *wrong direction*.

The classic example of that backward intuition was my own introduction to systems analysis, the World model. Asked by the Club of Rome—an international group of businessmen, statesmen, and scientists—to show how major global problems of poverty and hunger, environmental destruction, resource depletion, urban deterioration, and unemployment are related and how they might be solved, Forrester made a computer model and came out with a clear leverage point: growth.[2] Not only population growth, but economic growth. Growth has costs as well as benefits, and we typically don't count the costs—among which are poverty and hunger, environmental destruction, and so on—the whole list of problems we are trying to solve with growth! What is needed is much slower growth, very different kinds of growth, and in some cases no growth or negative growth.

The world's leaders are correctly fixated on economic growth as the answer to virtually all problems, *but they're pushing with all their might in the wrong direction.*

Another of Forrester's classics was his study of urban dynamics, published in 1969, which demonstrated that subsidized low-income housing is a leverage point.[3] The less of it there is, the *better off* the city is—even the low-income folks in the city. This model came out at a time when national policy dictated massive low-income housing projects, and Forrester was derided. Since then, many of those projects have been torn down in city after city.

Counterintuitive—that's Forrester's word to describe complex systems. Leverage points frequently are not intuitive. Or if they are, we too often use them backward, systematically worsening whatever problems we are trying to solve.

I have come up with no quick or easy formulas for finding leverage points in complex and dynamic systems. Give me a few months or years and I'll figure it out. And I know from bitter experience that, because they are so counterintuitive, when I do discover a system's leverage points, hardly anybody will believe me. Very frustrating—especially for those of us who yearn not just to understand complex systems, but to make the world work better.

It was in just such a moment of frustration that I proposed a list of places to intervene in a system during a meeting on the implications of global-trade regimes. I offer this list to you with much humility and wanting to leave room for its evolution. What bubbled up in me that day was distilled from decades of rigorous analysis of many different kinds of systems done by many smart people. But complex systems are, well, complex. It's dangerous to generalize about them. What you read here is still a work in progress; it's not a recipe for finding leverage points. Rather, it's an invitation to think more broadly about system change.

As systems become complex, their behavior can become surprising. Think about your checking account. You write checks and make deposits. A little interest keeps flowing in (if you have a large enough balance) and bank fees flow out even if you have no money in the account, thereby creating an accumulation of debt. Now attach your account to a thousand others and let the bank create loans as a function of your combined and fluctuating deposits, link a thousand of those banks into a federal reserve system—and you begin to see how simple stocks and flows, plumbed together, create systems way too complicated and dynamically complex to figure out easily.

That's why leverage points are often not intuitive. And that's enough systems theory to proceed to the list.

12. Numbers—Constants and parameters such as subsidies, taxes, standards

Think about the basic stock-and-flow bathtub from Chapter One. The size of the flows is a matter of numbers and how quickly those numbers can be changed. Maybe the faucet turns hard, so it takes a while to get the water flowing or to turn it off. Maybe the drain is blocked and can allow only a small flow, no matter how open it is. Maybe the faucet can deliver with the force of a fire hose. Some of these kinds of parameters are physically locked in and unchangeable, but many can be varied and so are popular intervention points.

Consider the national debt. It may seem like a strange stock; it is a money hole. The rate at which the hole deepens is called the annual deficit. Income from taxes shrinks the hole, government expenditures expand it. Congress and the president spend most of their time arguing about the many, many parameters that increase (spending) and decrease (taxing) the size or depth of the hole. Since those flows are connected to us, the voters, these are politically charged parameters. But, despite all the fireworks, and no matter which party is in charge, the money hole has been deepening for years now, just at different rates.

To adjust the dirtiness of the air we breathe, the government sets parameters called ambient-air-quality standards. To ensure some standing stock of forest (or some flow of money to logging companies), it sets allowed annual cuts. Corporations adjust parameters such as wage rates and product prices, with an eye on the level in their profit bathtub—the bottom line.

The amount of land we set aside for conservation each year. The minimum wage. How much we spend on AIDS research or Stealth bombers. The service charge the bank extracts from your account. All of these are parameters, adjustments to faucets. So, by the way, is firing people and getting new ones, including politicians. Putting different hands on the faucets may change the rate at which the faucets turn, but if they're the same old faucets, plumbed into the same old system, turned according to the same old information and goals and rules, the system behavior isn't going to change much. Electing Bill Clinton was definitely different from electing the elder George Bush, but not all that different, given that every president is plugged into the same political system. (Changing the way money flows in that system would make much more of a difference—but I'm getting ahead of myself on this list.)

Numbers, the sizes of flows, are dead last on my list of powerful interventions. Diddling with the details, arranging the deck chairs on the Titanic. Probably 90—no 95, no 99 percent—of our attention goes to parameters, but there's not a lot of leverage in them.

It's not that parameters aren't important—they can be, especially in the short term and to the individual who's standing directly in the flow. People care deeply about such variables as taxes and the minimum wage, and so fight fierce battles over them. But changing these variables *rarely changes the behavior of the national economy system*. If the system is chronically stagnant, parameter changes rarely kick-start it. If it's wildly variable, they usually don't stabilize it. If it's growing out of control, they don't slow it down.

Whatever cap we put on campaign contributions, it doesn't clean up politics. The Fed's fiddling with the interest rate hasn't made business cycles go away. (We always forget that during upturns, and are shocked, shocked by the downturns.) After decades of the strictest air pollution standards in the world, Los Angeles air is less dirty, but it isn't clean. Spending more on police doesn't make crime go away.

Since I'm about to get into some examples where parameters are leverage points, let me stick in a big caveat here. Parameters become leverage points when they go into ranges that kick off one of the items higher on this list. Interest rates, for example, or birth rates, control the gains around reinforcing feedback loops. System goals are parameters that can make big differences.

These kinds of critical numbers are not nearly as common as people seem to think they are. Most systems have evolved or are designed to stay far out of range of critical parameters. Mostly, the numbers are not worth the sweat put into them.

Here's a story a friend sent me over the Internet to makes that point:

> When I became a landlord, I spent a lot of time and energy trying to figure out what would be a "fair" rent to charge.
>
> I tried to consider all the variables, including the relative incomes of my tenants, my own income and cash-flow needs, which expenses were for upkeep and which were capital expenses, the equity versus the interest portion of the mortgage payments, how much my labor on the house was worth, etc.
>
> I got absolutely nowhere. Finally I went to someone who specializes in giving money advice. She said: "You're acting as though there is a fine line at which the rent is fair, and at any point above that point the tenant is being screwed and at any point below that you are being screwed. In fact, there is a large gray area in which both you and the tenant are getting a good, or at least a fair, deal. Stop worrying and get on with your life."[4]

11. Buffers—The sizes of stabilizing stocks relative to their flows

Consider a huge bathtub with slow in- and outflows. Now think about a small one with very fast flows. That's the difference between a lake and a river. You hear about catastrophic river floods much more often than catastrophic lake floods, because stocks that are big, relative to their flows, are more stable than small ones. In chemistry and other fields, a big, stabilizing stock is known as a buffer.

The stabilizing power of buffers is why you keep money in the bank rather than living from the flow of change through your pocket. It's why stores hold inventory instead of calling for new stock just as customers carry the old stock out the door. It's why we need to maintain more than the minimum breeding population of an endangered species. Soils in the eastern United States are more sensitive to acid rain than soils in the west, because they haven't got big buffers of calcium to neutralize acid.

You can often stabilize a system by increasing the capacity of a buffer.[5] But if a buffer is too big, the system gets inflexible. It reacts too slowly. And big buffers of some sorts, such as water reservoirs or inventories, cost a lot to build or maintain. Businesses invented just-in-time inventories, because occasional vulnerability to fluctuations or screw-ups is cheaper (for them, anyway) than certain, constant inventory costs—and because small-to-vanishing inventories allow more flexible response to shifting demand.

There's leverage, sometimes magical, in changing the size of buffers. But buffers are usually physical entities, not easy to change. The acid absorption capacity of eastern soils is not a leverage point for alleviating acid rain damage. The storage capacity of a dam is literally cast in concrete. So I haven't put buffers very high on the list of leverage points.

10. Stock-and-Flow Structures—Physical systems and their nodes of intersection

The plumbing structure, the stocks and flows and their physical arrangement, can have an enormous effect on how the system operates. When the Hungarian road system was laid out so all traffic from one side of the nation to the other had to pass through central Budapest, that determined a lot about air pollution and commuting delays that are not easily fixed by pollution control devices, traffic lights, or speed limits.

The only way to fix a system that is laid out poorly is to rebuild it, if you can. Amory Lovins and his team at Rocky Mountain Institute have done wonders on energy conservation by simply straightening out bent pipes and enlarging ones that are too small. If we did similar energy retrofits on all the buildings in the United States, we could shut down many of our electric power plants.

But often physical rebuilding is the slowest and most expensive kind of change to make in a system. Some stock-and-flow structures are just plain unchangeable. The baby-boom swell in the U.S. population first caused pressure on the elementary school system, then high schools, then colleges, then jobs and housing, and now we're supporting its retirement. There's not much we can do about it, because five-year-olds become six-year-olds, and sixty-four-year-olds become sixty-five-year-olds predictably and unstoppably. The same can be said for the lifetime of destructive CFC molecules in the ozone layer, for the rate at which contaminants get washed out of aquifers, for the fact that an inefficient car fleet takes ten to twenty years to turn over.

Physical structure is crucial in a system, but is rarely a leverage point, because changing it is rarely quick or simple. The leverage point is in proper design in the first place. After the structure is built, the leverage is in understanding its limitations and bottlenecks, using it with maximum efficiency, and refraining from fluctuations or expansions that strain its capacity.

9. Delays—The lengths of time relative to the rates of system changes

Delays in feedback loops are critical determinants of system behavior. They are common causes of oscillations. If you're trying to adjust a stock (your store inventory) to meet your goal, but you receive only delayed information about what the state of the stock is, you will overshoot and undershoot your goal. The same is true if your information is timely, but your response isn't. For example, it takes several years to build an electric power plant that will likely last thirty years. Those delays make it impossible to build exactly the right number of power plants to supply rapidly changing demand for electricity. Even with immense effort at forecasting, almost every electricity industry in the world experiences long oscillations between overcapacity and undercapacity. A system just can't respond to short-term changes when it has long-term delays. That's why a massive central-planning system, such as the Soviet Union or General Motors, necessarily functions poorly.

Because we know they're important, we see delays wherever we look. For example, the delay between the time when a pollutant is dumped on the land and when it trickles down to the groundwater; or the delay between the birth of a child and the time when that child is ready to have a child; or the delay between the first successful test of a new technology and the time when that technology is installed throughout the economy; or the time it takes for a price to adjust to a supply-demand imbalance.

A delay in a feedback process is critical *relative to rates of change in the stocks that the feedback loop is trying to control*. Delays that are too short cause overreaction, "chasing your tail," oscillations amplified by the jumpiness of the response. Delays that are too long cause damped, sustained, or exploding oscillations, depending on how much too long. Overlong delays in a system with a threshold, a danger point, a range past which irreversible damage can occur, cause overshoot and collapse.

I would list delay length as a high leverage point, except for the fact that delays are not often easily changeable. Things take as long as they take. You can't do a lot about the construction time of a major piece of capital, or the maturation time of a child, or the growth rate of a forest. It's usually easier *to slow down the change rate*, so that inevitable feedback delays won't cause so much trouble. That's why growth rates are higher up on the leverage-point list than delay times.

And that's why slowing economic growth is a greater leverage point in Forrester's World model than faster technological development or freer market prices. Those are attempts to speed up the rate of adjustment. But the world's physical capital stock, its factories and boilers, the concrete manifestations of its working technologies, can change only so fast, even in the face of new prices or new ideas—and prices and ideas don't change instantly either, not through a whole global culture. There's more leverage in slowing the system down so technologies and prices can keep up with it, than there is in wishing the delays would go away.

But if there is a delay in your system that *can* be changed, changing it can have big effects. Watch out! Be sure you change it in the right direction! (For example, the great push to reduce information and money-transfer delays in financial markets is just asking for wild gyrations.)

8. Balancing Feedback Loops—The strength of the feedbacks relative to the impacts they are trying to correct

Now we're beginning to move from the physical part of the system to the information and control parts, where more leverage can be found.

Balancing feedback loops are ubiquitous in systems. Nature evolves them and humans invent them as controls to keep important stocks within safe bounds. A thermostat loop is the classic example. Its purpose is to keep the system stock called "temperature of the room" fairly constant near a desired level. Any balancing feedback loop needs a goal (the thermostat setting), a monitoring and signaling device to detect deviation from the goal (the thermostat), and a response mechanism (the furnace and/or air conditioner, fans, pumps, pipes, fuel, etc.).

A complex system usually has numerous balancing feedback loops it can bring into play, so it can self-correct under different conditions and impacts. Some of those loops may be inactive much of the time—like the emergency cooling system in a nuclear power plant, or your ability to sweat or shiver to maintain your body temperature—but their presence is critical to the long-term welfare of the system.

One of the big mistakes we make is to strip away these "emergency" response mechanisms because they aren't often used and they appear to be costly. In the short term, we see no effect from doing this. In the long term, we drastically narrow the range of conditions over which the system can survive. One of the most heartbreaking ways we do this is in encroaching on the habitats of endangered species. Another is in encroaching on our own time for personal rest, recreation, socialization, and meditation.

The strength of a balancing loop—its ability to keep its appointed stock at or near its goal—depends on the combination of all its parameters and links—the accuracy and rapidity of monitoring, the quickness and power of response, the directness and size of corrective flows. Sometimes there are leverage points here.

Take markets, for example, the balancing feedback systems that are all but worshipped by many economists. They can indeed be marvels of self-correction, as prices vary to moderate supply and demand and keep them in balance. Price is the central piece of information signaling both producers and consumers. The more the price is kept clear, unambiguous, timely, and truthful, the more smoothly markets will operate. Prices that reflect *full* costs will tell consumers how much they can actually afford and will reward efficient producers. Companies and governments are fatally attracted to the price leverage point, but too often determinedly push it in the wrong direction with subsidies, taxes, and other forms of confusion.

These modifications weaken the feedback power of market signals by twisting information in their favor. The *real* leverage here is to keep them from doing it. Hence, the necessity of antitrust laws, truth-in-advertising laws, attempts to internalize costs (such as pollution fees), the removal of perverse subsidies, and other ways of leveling market playing fields.

Strengthening and clarifying market signals, such as full-cost accounting, don't get far these days, because of the weakening of another set of balancing feedback loops—those of democracy. This great system was invented to put self-correcting feedback between the people and their government. The people, informed about what their elected representatives do, respond by voting those representatives in or out of office. The process depends on the free, full, unbiased flow of information back and forth between electorate and leaders. Billions of dollars are spent to limit and bias and dominate that flow of clear information. Give the people who want to distort market-price signals the power to influence government leaders, allow the distributors of information to be self-interested partners, and none of the necessary balancing feedbacks work well. Both market and democracy erode.

The strength of a balancing feedback loop is important *relative to the impact it is designed to correct.* If the impact increases in strength, the feedbacks have to be strengthened too. A thermostat system may work fine on a cold winter day—but open all the windows and its corrective power is no match for the temperature change imposed on the system. Democracy works better without the brainwashing power of centralized mass communications. Traditional controls on fishing were sufficient until sonar spotting and drift nets and other technologies made it possible for a few actors to catch the last fish. The power of big industry calls for the power of big government to hold it in check; a global economy makes global regulations necessary.

Examples of strengthening balancing feedback controls to improve a system's self-correcting abilities include:

  • preventive medicine, exercise, and good nutrition to bolster
    the body's ability to fight disease,

- integrated pest management to encourage natural predators of crop pests,
- the Freedom of Information Act to reduce government secrecy,
- monitoring systems to report on environmental damage,
- protection for whistleblowers, and
- impact fees, pollution taxes, and performance bonds to recapture the externalized public costs of private benefits.

7. Reinforcing Feedback Loops—The strength of the gain of driving loops

A balancing feedback loop is self-correcting; a reinforcing feedback loop is self-reinforcing. The more it works, the more it gains power to work some more, driving system behavior in one direction. The more people catch the flu, the more they infect other people. The more babies are born, the more people grow up to have babies. The more money you have in the bank, the more interest you earn, the more money you have in the bank. The more the soil erodes, the less vegetation it can support, the fewer roots and leaves to soften rain and runoff, the more soil erodes. The more high-energy neutrons in the critical mass, the more they knock into nuclei and generate more high-energy neutrons, leading to a nuclear explosion or meltdown.

Reinforcing feedback loops are sources of growth, explosion, erosion, and collapse in systems. A system with an unchecked reinforcing loop ultimately will destroy itself. That's why there are so few of them. Usually a balancing loop will kick in sooner or later. The epidemic will run out of infectible people—or people will take increasingly stronger steps to avoid being infected. The death rate will rise to equal the birth rate—or people will see the consequences of unchecked population growth and have fewer babies. The soil will erode away to bedrock, and after a million years the bedrock will crumble into new soil—or people will stop overgrazing, put up check dams, plant trees, and stop the erosion.

In all those examples, the first outcome is what will happen if the reinforcing loop runs its course, the second is what will happen if there's an intervention to reduce its self-multiplying power. Reducing the gain around a reinforcing loop—slowing the growth—is usually a more powerful leverage point in systems than strengthening balancing loops, and far more preferable than letting the reinforcing loop run.

Population and economic growth rates in the World model are leverage points, because slowing them gives the many balancing loops, through technology and markets and other forms of adaptation (all of which have limits and delays), time to function. It's the same as slowing the car when you're driving too fast, rather than calling for more responsive brakes or technical advances in steering.

There are many reinforcing feedback loops in society that reward the winners of a competition with the resources to win even bigger next time—the "success to the successful" trap. Rich people collect interest; poor people pay it. Rich people pay accountants and lean on politicians to reduce their taxes; poor people can't. Rich people give their kids inheritances and good educations. Antipoverty programs are weak balancing loops that try to counter these strong reinforcing ones. It would be much more effective to weaken the reinforcing loops. That's what progressive income tax, inheritance tax, and universal high-quality public education programs are meant to do. If the wealthy can influence government to weaken, rather than strengthen, those measures, then the government itself shifts from a balancing structure to one that reinforces success to the successful!

Look for leverage points around birth rates, interest rates, erosion rates, "success to the successful" loops, any place where the more you have of something, the more you have the possibility of having more.

6. Information Flows—The structure of who does and does not have access to information

In Chapter Four, we examined the story of the electric meter in a Dutch housing development—in some of the houses the meter was installed in the basement; in others it was installed in the front hall. With no other differences in the houses, electricity consumption was 30 percent lower in the houses where the meter was in the highly visible location in the front hall.

I love that story because it's an example of a high leverage point in the information structure of the system. It's not a parameter adjustment, not a strengthening or weakening of an existing feedback loop. It's a new loop, delivering feedback to a place where it wasn't going before.

Missing information flows is one of the most common causes of system malfunction. Adding or restoring information can be a powerful intervention, usually much easier and cheaper than rebuilding physical infrastructure. The tragedy of the commons that is crashing the world's commercial fisheries occurs because there is little feedback from the state of the fish population to the decision to invest in fishing vessels. Contrary to economic opinion, the *price* of fish doesn't provide that feedback. As the fish get more scarce they become more expensive, and it becomes all the more profitable to go out and catch the last few. That's a perverse feedback, a reinforcing loop that leads to collapse. It is not price information but population information that is needed.

It's important that the missing feedback be restored to the right place and in compelling form. To take another tragedy of the commons example, it's not enough to inform all the users of an aquifer that the groundwater level is dropping. That could initiate a race to the bottom. It would be more effective to set the cost of water to rise steeply as the pumping rate begins to exceed the recharge rate.

Other examples of compelling feedback are not hard to find. Suppose taxpayers got to specify on their return forms what government services their tax payments must be spent on. (Radical democracy!) Suppose any town or company that puts a water intake pipe in a river had to put it immediately *downstream* from its own wastewater outflow pipe. Suppose any public or private official who made the decision to invest in a nuclear power plant got the waste from that facility stored on his or her lawn. Suppose (this is an old one) the politicians who declare war were required to spend that war in the front lines.

There is a systematic tendency on the part of human beings to avoid accountability for their own decisions. That's why there are so many missing feedback loops—and why this kind of leverage point is so often popular with the masses, unpopular with the powers that be, and effective, if you can get the powers that be to permit it to happen (or go around them and make it happen anyway).

5. Rules—Incentives, punishments, constraints

The rules of the system define its scope, its boundaries, its degrees of freedom. Thou shalt not kill. Everyone has the right of free speech. Contracts are to be honored. The president serves four-year terms and cannot serve more than two of them. Nine people on a team, you have to touch every base, three strikes and you're out. If you get caught robbing a bank, you go to jail.

Mikhail Gorbachev came to power in the Soviet Union and opened information flows (glasnost) and changed the economic rules (perestroika), and the Soviet Union saw tremendous change.

Constitutions are the strongest examples of social rules. Physical laws such as the second law of thermodynamics are absolute rules, whether we understand them or not or like them or not. Laws, punishments, incentives, and informal social agreements are progressively weaker rules.

To demonstrate the power of rules, I like to ask my students to imagine different ones for a college. Suppose the students graded the teachers, or each other. Suppose there were no degrees: You come to college when you want to learn something, and you leave when you've learned it. Suppose tenure were awarded to professors according to their ability to solve real-world problems, rather than to publish academic papers. Suppose a class got graded as a group, instead of as individuals.

As we try to imagine restructured rules and what our behavior would be under them, we come to understand the power of rules. They are high leverage points. Power over the rules is real power. That's why lobbyists congregate when Congress writes laws, and why the Supreme Court, which interprets and delineates the Constitution—the rules for writing the rules—has even more power than Congress. If you want to understand the deepest malfunctions of systems, pay attention to the rules and to who has power over them.

That's why my systems intuition was sending off alarm bells as the new world trade system was explained to me. It is a system with rules designed by corporations, run by corporations, for the benefit of corporations. Its rules exclude almost any feedback from any other sector of society. Most of its meetings are closed even to the press (no information flow, no feedback). It forces nations into reinforcing loops "racing to the bottom," competing with each other to weaken environmental and social safeguards in order to attract corporate investment. It's a recipe for unleashing "success to the successful" loops, until they generate enormous accumulations of power and huge centralized planning systems that will destroy themselves.

4. Self-Organization—The power to add, change, or evolve system structure

The most stunning thing living systems and some social systems can do is to change themselves utterly by creating whole new structures and behaviors. In biological systems that power is called evolution. In human economies it's called technical advance or social revolution. In systems lingo it's called self-organization.

Self-organization means changing any aspect of a system lower on this list—adding completely new physical structures, such as brains or wings or computers—adding new balancing or reinforcing loops, or new rules. The ability to self-organize is the strongest form of system resilience. A system that can evolve can survive almost any change, by changing itself. The human immune system has the power to develop new responses to some kinds of insults it has never before encountered. The human brain can take in new information and pop out completely new thoughts.

The power of self-organization seems so wondrous that we tend to regard it as mysterious, miraculous, heaven sent. Economists often model technology as magic—coming from nowhere, costing nothing, increasing the productivity of an economy by some steady percent each year. For centuries people have regarded the spectacular variety of nature with the same awe. Only a divine creator could bring forth such a creation.

Further investigation of self-organizing systems reveals that the divine creator, if there is one, does not have to produce evolutionary miracles. He, she, or it just has to write marvelously clever *rules for self-organization*. These rules basically govern how, where, and what the system can add onto or subtract from itself under what conditions. As hundreds of self-organizing computer models have demonstrated, complex and delightful patterns can evolve from quite simple sets of rules. The genetic code within the DNA that is the basis of all biological evolution contains just four different letters, combined into words of three letters each. That pattern, and the rules for replicating and rearranging it, has been constant for something like three billion years, during which it has spewed out an unimaginable variety of failed and successful self-evolved creatures.

Self-organization is basically a matter of an evolutionary raw material—a highly variable stock of information from which to select possible patterns—and a means for experimentation, for selecting and testing new patterns. For biological evolution, the raw material is DNA, one source of variety is spontaneous mutation, and the testing mechanism is a changing environment in which some individuals do not survive to reproduce. For technology, the raw material is the body of understanding science has accumulated and stored in libraries and in the brains of its practitioners. The source of variety is human creativity (whatever *that* is) and the selection mechanism can be whatever the market will reward, or whatever governments and foundations will fund, or whatever meets human needs.

When you understand the power of system self-organization, you begin to understand why biologists worship biodiversity even more than economists worship technology. The wildly varied stock of DNA, evolved and accumulated over billions of years, is the source of evolutionary potential, just as science libraries and labs and universities where scientists are trained are the source of technological potential. Allowing species to go extinct is a systems crime, just as randomly eliminating all copies of particular science journals or particular kinds of scientists would be.

The same could be said of human cultures, of course, which are the store of behavioral repertoires, accumulated over not billions, but hundreds of thousands of years. They are a stock out of which social evolution can arise. Unfortunately, people appreciate the precious evolutionary potential of cultures even less than they understand the preciousness of every genetic variation in the world's ground squirrels. I guess that's because one aspect of almost every culture is the belief in the utter superiority of that culture.

Insistence on a single culture shuts down learning and cuts back resilience. Any system, biological, economic, or social, that gets so encrusted that it cannot self-evolve, a system that systematically scorns experimentation and wipes out the raw material of innovation, is doomed over the long term on this highly variable planet.

The intervention point here is obvious, but unpopular. Encouraging variability and experimentation and diversity means "losing control." Let a thousand flowers bloom and *anything* could happen! Who wants that?

Let's play it safe and push this lever in the wrong direction by wiping out biological, cultural, social, and market diversity!

3. Goals—The purpose or function of the system

Right there, the diversity-destroying consequence of the push for control demonstrates why the goal of a system is a leverage point superior to the self-organizing ability of a system. If the goal is to bring more and more of the world under the control of one particular central planning system (the empire of Genghis Khan, the Church, the People's Republic of China, Wal-Mart, Disney), then everything further down the list, physical stocks and flows, feedback loops, information flows, even self-organizing behavior, will be twisted to conform to that goal.

That's why I can't get into arguments about whether genetic engineering is a "good" or a "bad" thing. Like all technologies, it depends on who is wielding it, with what goal. The only thing one can say is that if corporations wield it for the purpose of generating marketable products, that is a very different goal, a very different selection mechanism, a very different direction for evolution than anything the planet has seen so far.

As my little single-loop examples have shown, most balancing feedback loops within systems have their own goals—to keep the bathwater at the right level, to keep the room temperature comfortable, to keep inventories stocked at sufficient levels, to keep enough water behind the dam. Those goals are important leverage points for pieces of systems, and most people realize that. If you want the room warmer, you know the thermostat setting is the place to intervene. But there are larger, less obvious, higher-leverage goals, those of the entire system.

Even people within systems don't often recognize what whole-system goal they are serving. "To make profits," most corporations would say, but that's just a rule, a necessary condition to stay in the game. What is the point of the game? To grow, to increase market share, to bring the world (customers, suppliers, regulators) more and more under the control of the corporation, so that its operations becomes ever more shielded from uncertainty. John Kenneth Galbraith recognized that corporate goal—to engulf everything—long ago.[6] It's the goal of a cancer too. Actually it's the goal of every living population—and only a bad one when it isn't balanced by higher-level balancing feedback loops that never let an upstart power-loop-driven entity control the world. The goal of keeping the market competitive has to trump the goal of each individual corporation to eliminate its competitors, just as in ecosystems, the goal of keeping populations in balance and evolving has to trump the goal of each population to reproduce without limit.

I said a while back that changing the players in the system is a low-level intervention, as long as the players fit into the same old system. The exception to that rule is at the top, where a single player can have the power to change the system's goal. I have watched in wonder as—only very occasionally—a new leader in an organization, from Dartmouth College to Nazi Germany, comes in, enunciates a new goal, and swings hundreds or thousands or millions of perfectly intelligent, rational people off in a new direction.

That's what Ronald Reagan did, and we watched it happen. Not long before he came to office, a president could say "Ask not what government can do for you, ask what you can do for the government," and no one even laughed. Reagan said over and over, the goal is not to get the people to help the government and not to get government to help the people, but to get government off our backs. One can argue, and I would, that larger system changes and the rise of corporate power over government let him get away with that. But the thoroughness with which the public discourse in the United States and even the world has been changed since Reagan is testimony to the high leverage of articulating, meaning, repeating, standing up for, insisting upon, new system goals.

2. Paradigms—The mind-set out of which the system—its goals, structure, rules, delays, parameters—arises

Another of Jay Forrester's famous systems sayings goes: It doesn't matter how the tax law of a country is written. There is a shared idea in the minds of the society about what a "fair" distribution of the tax load is. Whatever the laws say, by fair means or foul, by complications, cheating, exemptions or deductions, by constant sniping at the rules, actual tax payments will push right up against the accepted idea of "fairness."

The shared idea in the minds of society, the great big unstated assumptions, constitute that society's paradigm, or deepest set of beliefs about how the world works. These beliefs are unstated because it is unnecessary to state them—everyone already knows them. Money measures something real and has real meaning; therefore, people who are paid less are literally worth less. Growth is good. Nature is a stock of resources to be converted to human purposes. Evolution stopped with the emergence of *Homo sapiens*. One can "own" land. Those are just a few of the paradigmatic assumptions of our current culture, all of which have utterly dumbfounded other cultures, who thought them not the least bit obvious.

Paradigms are the sources of systems. From them, from shared social agreements about the nature of reality, come system goals and information flows, feedbacks, stocks, flows, and everything else about systems. No one has ever said that better than Ralph Waldo Emerson:

> Every nation and every man instantly surround themselves with a material apparatus which exactly corresponds to . . . their state of thought. Observe how every truth and every error, each a thought of some man's mind, clothes itself with societies, houses, cities, language, ceremonies, newspapers. Observe the ideas of the present day . . . see how timber, brick, lime, and stone have flown into convenient shape, obedient to the master idea reigning in the minds of many persons. . . . It follows, of course, that the least enlargement of ideas . . . would cause the most striking changes of external things.[7]

The ancient Egyptians built pyramids because they believed in an afterlife. We build skyscrapers because we believe that space in downtown cities is enormously valuable. Whether it was Copernicus and Kepler showing that the earth is not the center of the universe, or Einstein hypothesizing that matter and energy are interchangeable, or Adam Smith postulating that the selfish actions of individual players in markets wonderfully accumulate to the common good, people who have managed to intervene in systems at the level of paradigm have hit a leverage point that totally transforms systems.

You could say paradigms are harder to change than anything else about a system, and therefore this item should be lowest on the list, not second-to-highest. But there's nothing physical or expensive or even slow in the process of paradigm change. In a single individual it can happen in a millisecond. All it takes is a click in the mind, a falling of scales from the eyes, a new way of seeing. Whole societies are another matter—they resist challenges to their paradigms harder than they resist anything else.

So how do you change paradigms? Thomas Kuhn, who wrote the seminal book about the great paradigm shifts of science, has a lot to say about that.[8] You keep pointing at the anomalies and failures in the old paradigm. You keep speaking and acting, loudly and with assurance, from the new one. You insert people with the new paradigm in places of public visibility and power. You don't waste time with reactionaries; rather, you work with active change agents and with the vast middle ground of people who are open-minded.

Systems modelers say that we change paradigms by building a model of the system, which takes us outside the system and forces us to see it whole. I say that because my own paradigms have been changed that way.

1. Transcending Paradigms

There is yet one leverage point that is even higher than changing a paradigm. That is to keep oneself unattached in the arena of paradigms, to stay flexible, to realize that *no* paradigm is "true," that every one, including the one that sweetly shapes your own worldview, is a tremendously limited understanding of an immense and amazing universe that is far beyond human comprehension. It is to "get" at a gut level the paradigm that there are paradigms, and to see that that itself is a paradigm, and to regard that whole realization as devastatingly funny. It is to let go into not-knowing, into what the Buddhists call enlightenment.

People who cling to paradigms (which means just about all of us) take one look at the spacious possibility that everything they think is guaranteed to be nonsense and pedal rapidly in the opposite direction. Surely there is no power, no control, no understanding, not even a reason for being, much less acting, embodied in the notion that there is no certainty in *any* worldview. But, in fact, everyone who has managed to entertain that idea, for a moment or for a lifetime, has found it to be the basis for radical empowerment. If no paradigm is right, you can choose whatever one will help to achieve your purpose. If you have no idea where to get a purpose, you can listen to the universe.

It is in this space of mastery over paradigms that people throw off addictions, live in constant joy, bring down empires, get locked up or burned at the stake or crucified or shot, and have impacts that last for millennia.

There is so much that could be said to qualify this list of places to intervene in a system. It is a tentative list and its order is slithery. There are exceptions to every item that can move it up or down the order of leverage. Having had the list percolating in my subconscious for years has not transformed me into Superwoman. The higher the leverage point, the more the system will resist changing it—that's why societies often rub out truly enlightened beings.

Magical leverage points are not easily accessible, even if we know where they are and which direction to push on them. There are no cheap tickets to mastery. You have to work hard at it, whether that means rigorously analyzing a system or rigorously casting off your own paradigms and throwing yourself into the humility of not-knowing. In the end, it seems that mastery has less to do with pushing leverage points than it does with strategically, profoundly, madly, letting go and dancing with the system.

SEVEN

Living in a World of Systems

The real trouble with this world of ours is not that it is an unreasonable world, nor even that it is a reasonable one. The commonest kind of trouble is that it is nearly reasonable, but not quite. Life is not an illogicality; yet it is a trap for logicians. It looks just a little more mathematical and regular than it is.

—G. K. Chesterton,[1] 20th century writer

People who are raised in the industrial world and who get enthused about systems thinking are likely to make a terrible mistake. They are likely to assume that here, in systems analysis, in interconnection and complication, in the power of the computer, here at last, is the key to prediction and control. This mistake is likely because the mind-set of the industrial world assumes that there is a key to prediction and control.

I assumed that at first too. We all assumed it, as eager systems students at the great institution called MIT. More or less innocently, enchanted by what we could see through our new lens, we did what many discoverers do. We exaggerated our findings. We did so not with any intent to deceive others, but in the expression of our own expectations and hopes. Systems thinking for us was more than subtle, complicated mind play. It was going to *make systems work*.

Like the explorers searching for the passage to India who ran into the Western Hemisphere instead, we had found something, but it wasn't what we thought we had found. It was something so different from what we had been looking for that we didn't know what to make of it. As we got to know systems thinking better, it turned out to have greater worth than we had thought, but not in the way we had thought.

Our first comeuppance came as we learned that it's one thing to understand how to fix a system and quite another to wade in and fix it. We had many earnest discussions on the topic of "implementation," by which we meant "how to get managers and mayors and agency heads to follow our advice."

The truth was, *we* didn't even follow our advice. We gave learned lectures on the structure of addiction and could not give up coffee. We knew all about the dynamics of eroding goals and eroded our own jogging programs. We warned against the traps of escalation and shifting the burden and then created them in our own marriages.

Social systems are the external manifestations of cultural thinking patterns and of profound human needs, emotions, strengths, and weaknesses. Changing them is not as simple as saying "now all change," or of trusting that he who knows the good shall do the good.

We ran into another problem. Our systems insights helped us understand many things we hadn't understood before, but they didn't help us understand *everything*. In fact, they raised at least as many questions as they answered. Like all the other lenses humanity has developed with which to peer into macrocosms and microcosms, this one too revealed wondrous new things, many of which were wondrous new mysteries. The mysteries our new tool revealed lay especially within the human mind and heart and soul. Here are just few of the questions that were prompted by our insights into how systems work.

A systems insight . . . can raise more questions!

Systems thinkers are by no means the first or only people to ask questions like these. When we started asking them, we found whole disciplines, libraries, histories, asking the same questions, and to some extent offering answers. What was unique about our search was not our answers, or even our questions, but the fact that the tool of systems thinking, born out of engineering and mathematics, implemented in computers, drawn from a mechanistic mind-set and a quest for prediction and control, leads its practitioners, inexorably I believe, to confront the most deeply human mysteries. Systems thinking makes clear even to the most committed technocrat that getting along in this world of complex systems requires more than technocracy.

Self-organizing, nonlinear, feedback systems are inherently unpredictable. They are not controllable. They are understandable only in the most general way. The goal of foreseeing the future exactly and preparing for it

A new information feedback loop at *this* point in this system will make it behave much better. But the decision makers are resistant to the information they need! They don't pay attention to it, they don't believe it, they don't know how to interpret it.

If *this* feedback loop could just be oriented around *that* value, the system would produce a result that everyone wants. (Not more energy, but more energy services. Not GNP, but material sufficiency and security. Not growth, but progress.) We don't have to change anyone's values, we just have to get the system to operate around real values.

Here is a system that seems perverse on all counts. It produces inefficiency, ugliness, environmental degradation, and human misery. But if we sweep it away, we will have no system. Nothing is more frightening than that. (As I write, I have the former communist system of the Soviet Union in mind, but that is not the only possible example.)

The people in this system are putting up with deleterious behavior because they are afraid of change. They don't trust that a better system is possible. They feel they have no power to demand or bring about improvement.

perfectly is unrealizable. The idea of making a complex system do just what you want it to do can be achieved only temporarily, at best. We can never fully understand our world, not in the way our reductionist science has led us to expect. Our science itself, from quantum theory to the mathematics of chaos, leads us into irreducible uncertainty. For any objective other than the most trivial, we can't optimize; we don't even know what to optimize. We can't keep track of everything. We can't find a proper, sustainable relationship to nature, each other, or the institutions we create, if we try to do it from the role of omniscient conqueror.

For those who stake their identity on the role of omniscient conqueror,

Why do people actively sort and screen information the way they do? How do they determine what to let in and what to let bounce off, what to reckon with and what to ignore or disparage? How is it that, exposed to the same information, different people absorb different messages, and draw different conclusions?

What are values? Where do they come from? Are they universal, or culturally determined? What causes a person or a society to give up on attaining "real values" and to settle for cheap substitutes? How can you key a feedback loop to qualities you can't measure, rather than to quantities you can?

Why is it that periods of minimum structure and maximum freedom to create are so frightening? How is it that one way of seeing the world becomes so widely shared that institutions, technologies, production systems, buildings, cities, become shaped around that way of seeing? How do systems create cultures? How do cultures create systems? Once a culture and system have been found lacking, do they have to change through breakdown and chaos?

Why are people so easily convinced of their powerlessness? How do they become so cynical about their ability to achieve their visions? Why are they more likely to listen to people who tell them they can't make changes than they are to people who tell them they can?

the uncertainty exposed by systems thinking is hard to take. If you can't understand, predict, and control, what is there to do?

Systems thinking leads to another conclusion, however, waiting, shining, obvious, as soon as we stop being blinded by the illusion of control. It says that there is plenty to do, of a different sort of "doing." The future can't be predicted, but it can be envisioned and brought lovingly into being. Systems can't be controlled, but they can be designed and redesigned. We can't surge forward with certainty into a world of no surprises, but we can expect surprises and learn from them and even profit from them. We can't impose our will on a system. We can listen to what the system tells us, and discover how its properties and our values can work together to bring forth something much better than could ever be produced by our will alone.

We can't control systems or figure them out. But we can dance with them!

I already knew that, in a way. I had learned about dancing with great powers from whitewater kayaking, from gardening, from playing music, from skiing. All those endeavors require one to stay wide awake, pay close attention, participate flat out, and respond to feedback. It had never occurred to me that those same requirements might apply to intellectual work, to management, to government, to getting along with people.

But there it was, the message emerging from every computer model we made. Living successfully in a world of systems requires more of us than our ability to calculate. It requires our full humanity—our rationality, our ability to sort out truth from falsehood, our intuition, our compassion, our vision, and our morality.[2]

I want to end this chapter and this book by trying to summarize the most general "systems wisdoms" I have absorbed from modeling complex systems and from hanging out with modelers. These are the take-home lessons, the concepts and practices that penetrate the discipline of systems so deeply that one begins, however imperfectly, to practice them not just in one's profession, but in all of life. They are the behaviorial consequences of a worldview based on the ideas of feedback, nonlinearity, and systems responsible for their own behavior. When that engineering professor at Dartmouth noticed that we systems folks were "different" and wondered why, these, I think, were the differences he noticed.

The list probably isn't complete, because I am still a student in the school of systems. And it isn't a list that is unique to systems thinking; there are many ways to learn to dance. But here, as a start-off dancing lesson, are the practices I see my colleagues adopting, consciously or unconsciously, as they encounter new systems.

Get the Beat of the System

Before you disturb the system in any way, watch how it behaves. If it's a piece of music or a whitewater rapid or a fluctuation in a commodity price, study its beat. If it's a social system, watch it work. Learn its history. Ask people who've been around a long time to tell you what has happened. If possible, find or make a time graph of actual data from the system—peoples' memories are not always reliable when it comes to timing.

This guideline is deceptively simple. Until you make it a practice, you won't believe how many wrong turns it helps you avoid. Starting with the behavior of the system forces you to focus on facts, not theories. It keeps you from falling too quickly into your own beliefs or misconceptions, or those of others.

It's amazing how many misconceptions there can be. People will swear that rainfall is decreasing, say, but when you look at the data, you find that what is really happening is that variability is increasing—the droughts are deeper, but the floods are greater too. I have been told with great authority that the price of milk was going up when it was going down, that real interest rates were falling when they were rising, that the deficit was a higher fraction of the GNP than ever before when it wasn't.

It's especially interesting to watch how the various elements in the system do or do not vary together. Watching what really happens, instead of listening to peoples' theories of what happens, can explode many careless causal hypotheses. Every selectman in the state of New Hampshire seems to be positive that growth in a town will lower taxes, but if you plot growth rates against tax rates, you find a scatter as random as the stars in a New Hampshire winter sky. There is no discernible relationship at all.

Starting with the behavior of the system directs one's thoughts to dynamic, not static, analysis—not only to "What's wrong?" but also to "How did we get there?" "What other behavior modes are possible?" "If we don't change direction, where are we going to end up?" And looking to the strengths of the system, one can ask "What's working well here?" Starting with the history of several variables plotted together begins to suggest not only what elements are in the system, but how they might be interconnected.

And finally, starting with history discourages the common and distracting tendency we all have to define a problem not by the system's actual behavior, but by the lack of our favorite solution. (The problem is, we need to find more oil. The problem is, we need to ban abortion. The problem is, we don't have enough salesmen. The problem is, how can we attract more growth to this town?) Listen to any discussion, in your family or a committee meeting at work or among the pundits in the media, and watch people leap to solutions, usually solutions in "predict, control, or impose your will" mode, without having paid any attention to what the system is doing and why it's doing it.

Expose Your Mental Models to the Light of Day

When we draw structural diagrams and then write equations, we are forced to make our assumptions visible and to express them with rigor. We have to put every one of our assumptions about the system out where others (and we ourselves) can see them. Our models have to be complete, and they have to add up, and they have to be consistent. Our assumptions can no longer slide around (mental models are very slippery), assuming one thing for purposes of one discussion and something else contradictory for purposes of the next discussion.

You don't have to put forth your mental model with diagrams and equations, although doing so is a good practice. You can do it with words or lists or pictures or arrows showing what you think is connected to what. The more you do that, in any form, the clearer your thinking will become, the faster you will admit your uncertainties and correct your mistakes, and the more flexible you will learn to be. Mental flexibility—the willingness to redraw boundaries, to notice that a system has shifted into a new mode, to see how to redesign structure—is a necessity when you live in a world of flexible systems.

Remember, always, that everything you know, and everything everyone knows, is only a model. Get your model out there where it can be viewed. Invite others to challenge your assumptions and add their own. Instead of becoming a champion for one possible explanation or hypothesis or model, collect as many as possible. Consider all of them to be plausible until you find some evidence that causes you to rule one out. That way you will be emotionally able to see the evidence that rules out an assumption that may become entangled with your own identity.

Getting models out into the light of day, making them as rigorous as possible, testing them against the evidence, and being willing to scuttle them if they are no longer supported is nothing more than practicing the scientific method—something that is done too seldom even in science, and is done hardly at all in social science or management or government or everyday life.

Honor, Respect, and Distribute Information

You've seen how information holds systems together and how delayed, biased, scattered, or missing information can make feedback loops malfunction. Decision makers can't respond to information they don't have, can't respond accurately to information that is inaccurate, and can't respond in a timely way to information that is late. I would guess that most of what goes wrong in systems goes wrong because of biased, late, or missing information.

If I could, I would add an eleventh commandment to the first ten: *Thou shalt not distort, delay, or withhold information.* You can drive a system crazy by muddying its information streams. You can make a system work better with surprising ease if you can give it more timely, more accurate, more complete information.

For example, in 1986, new federal legislation, the Toxic Release Inventory, required U.S. companies to report all hazardous air pollutants emitted from each of their factories each year. Through the Freedom of Information Act (from a systems point of view, one of the most important laws in the nation), that information became a matter of public record. In July 1988, the first data on chemical emissions became available. The reported emissions were not illegal, but they didn't look very good when they were published in local papers by enterprising reporters, who had a tendency to make lists of "the top ten local polluters." That's all that happened. There were no lawsuits, no required reductions, no fines, no penalties. But within two years chemical emissions nationwide (at least as reported, and presumably also in fact) had decreased by 40 percent. Some companies were launching policies to bring their emissions down by 90 percent, just because of the release of previously withheld information.[3]

Information is power. Anyone interested in power grasps that idea very quickly. The media, the public relations people, the politicians, and advertisers who regulate much of the public flow of information have far more power than most people realize. They filter and channel information. Often they do so for short-term, self-interested purposes. It's no wonder our that social systems so often run amok.

Use Language with Care and Enrich It with Systems Concepts

Our information streams are composed primarily of language. Our mental models are mostly verbal. Honoring information means above all avoiding language pollution—making the cleanest possible use we can of language. Second, it means expanding our language so we can talk about complexity.

Fred Kofman wrote in a systems journal:

> [Language] can serve as a medium through which we create new understandings and new realities as we begin to talk about them. In fact, we don't talk about what we see; *we see only what we can talk about.* Our perspectives on the world depend on the interaction of our nervous system and our language—both act as filters through which we perceive our world. . . . The language and information systems of an organization are not an objective means of describing an outside reality—they fundamentally structure the perceptions and actions of its members. To reshape the measurement and communication systems of a [society] is to reshape all potential interactions at the most fundamental level. Language . . . as articulation of reality is more primordial than strategy, structure, or . . . culture.[4]

A society that talks incessantly about "productivity" but that hardly understands, much less uses, the word "resilience" is going to become productive and not resilient. A society that doesn't understand or use the term "carrying capacity" will exceed its carrying capacity. A society that talks about "creating jobs" as if that's something only companies can do will not inspire the great majority of its people to create jobs, for themselves or anyone else. Nor will it appreciate its workers for their role in "creating profits." And of course a society that talks about a "Peacekeeper" missile or "collateral damage," a "Final Solution" or "ethnic cleansing," is speaking what Wendell Berry calls "tyrannese."

> My impression is that we have seen, for perhaps a hundred and fifty years, a gradual increase in language that is either meaningless or destructive of meaning. And I believe that this increasing

unreliability of language parallels the increasing disintegration, over the same period, of persons and communities. . . .

He goes on to say:

> In this degenerative accounting, language is almost without the power of designation, because it is used conscientiously to refer to nothing in particular. Attention rests upon percentages, categories, abstract functions. . . . It is not language that the user will very likely be required to stand by or to act on, for it does not define any personal ground for standing or acting. Its only practical utility is to support with "expert opinion" a vast, impersonal technological action already begun. . . . It is a tyrannical language: tyrannese.[5]

The first step in respecting language is keeping it as concrete, meaningful, and truthful as possible—part of the job of keeping information streams clear. The second step is to enlarge language to make it consistent with our enlarged understanding of systems. If the Eskimos have so many words for snow, it's because they have studied and learned how to use snow. They have turned snow into a resource, a system with which they can dance. The industrial society is just beginning to have and use words for systems, because it is only beginning to pay attention to and use complexity. *Carrying capacity, structure, diversity*, and even *system* are old words that are coming to have richer and more precise meanings. New words are having to be invented.

My word processor has spell-check capability, which lets me add words that didn't originally come in its comprehensive dictionary. It's interesting to see what words I had to add when writing this book: *feedback, throughput, overshoot, self-organization, sustainability*.

Pay Attention to What Is Important, Not Just What Is Quantifiable

Our culture, obsessed with numbers, has given us the idea that what we can measure is more important than what we can't measure. Think about that for a minute. It means that we make quantity more important than quality. If quantity forms the goals of our feedback loops, if quantity is the center of our attention and language and institutions, if we motivate ourselves, rate ourselves, and reward ourselves on our ability to produce quantity, then quantity will be the result. You can look around and make up your own mind about whether quantity or quality is the outstanding characteristic of the world in which you live.

As modelers we have exposed ourselves to the ridicule of our scientific colleagues more than once by putting variables labeled "prejudice," or "self-esteem," or "quality of life" into our models. Since computers require numbers, we have had to make up quantitative scales by which to measure these qualitative concepts. "Let's say prejudice is measured from −10 to +10, where 0 means you are treated with no bias at all, −10 means extreme negative prejudice, and +10 means such positive prejudice that you can do no wrong. Now, suppose that you were treated with a prejudice of −2, or +5, or −8. What would that do to your performance at work?"

The relationship between prejudice and performance actually had to be put in a model once.[6] The study was for a firm that wanted to know how to do better at treating minority workers and how to move them up the corporate ladder. Everyone interviewed agreed that there certainly was a real connection between prejudice and performance. It was arbitrary what kind of scale to measure it by—it could have been 1 to 5 or 0 to 100—but it would have been much more unscientific to leave "prejudice" out of that study than to try to include it. When the workers in the company were asked to draw the relationship between their performance and prejudice, they came up with one of the most nonlinear relationships I've ever seen in a model.

Pretending that something doesn't exist if it's hard to quantify leads to faulty models. You've already seen the system trap that comes from setting goals around what is easily measured, rather than around what is important. So don't fall into that trap. Human beings have been endowed not only with the ability to count, but also with the ability to assess quality. Be a quality detector. Be a walking, noisy Geiger counter that registers the presence or absence of quality.

If something is ugly, say so. If it is tacky, inappropriate, out of proportion, unsustainable, morally degrading, ecologically impoverishing, or humanly demeaning, don't let it pass. Don't be stopped by the "if you can't define it and measure it, I don't have to pay attention to it" ploy. No one can define or measure justice, democracy, security, freedom, truth, or love. No one can define or measure any value. But if no one speaks up for them, if systems aren't designed to produce them, if we don't speak about them and point toward their presence or absence, they will cease to exist.

Make Feedback Policies for Feedback Systems

President Jimmy Carter had an unusual ability to think in feedback terms and to make feedback policies. Unfortunately, he had a hard time explaining them to a press and public that didn't understand feedback.

He suggested, at a time when oil imports were soaring, that there be a tax on gasoline proportional to the fraction of U.S. oil consumption that had to be imported. If imports continued to rise, the tax would rise until it suppressed demand and brought forth substitutes and reduced imports. If imports fell to zero, the tax would fall to zero.

The tax never got passed.

Carter also was trying to deal with a flood of illegal immigrants from Mexico. He suggested that nothing could be done about that immigration as long as there was a great gap in opportunity and living standards between the United States and Mexico. Rather than spending money on border guards and barriers, he said, we should spend money helping to build the Mexican economy, and we should continue to do so until the immigration stopped.

That never happened either.

You can imagine why a dynamic, self-adjusting feedback system cannot be governed by a static, unbending policy. It's easier, more effective, and usually much cheaper to design policies that change depending on the state of the system. Especially where there are great uncertainties, the best policies not only contain feedback loops, but meta-feedback loops—loops that alter, correct, and expand loops. These are policies that design *learning* into the management process.

An example was the historic Montreal Protocol to protect the ozone layer of the stratosphere. In 1987, when that protocol was signed, there was no certainty about the danger to the ozone layer, about the rate at which it was degrading, or about the specific effect of different chemicals. The protocol set targets for how fast the manufacture of the most damaging chemicals should be decreased. But it also required monitoring the situation and reconvening an international congress to change the phase-out schedule, if the damage to the ozone layer turned out to be more or less than expected. Just three years later, in 1990, the schedule had to be hurried forward and more chemicals added to it, because the damage was turning out to be much greater than was foreseen in 1987.

That was a feedback policy, structured for learning. We all hope that it worked in time.

Go for the Good of the Whole

Remember that hierarchies exist to serve the bottom layers, not the top. Don't maximize parts of systems or subsystems while ignoring the whole. Don't, as Kenneth Boulding once said, go to great trouble to optimize something that never should be done at all. Aim to enhance total systems properties, such as growth, stability, diversity, resilience, and sustainability—whether they are easily measured or not.

Listen to the Wisdom of the System

Aid and encourage the forces and structures that help the system run itself. Notice how many of those forces and structures are at the bottom of the hierarchy. Don't be an unthinking intervenor and destroy the system's own self-maintenance capacities. Before you charge in to make things better, pay attention to the value of what's already there.

A friend of mine, Nathan Gray, was once an aid worker in Guatemala. He told me of his frustration with agencies that would arrive with the intention of "creating jobs" and "increasing entrepreneurial abilities" and "attracting outside investors." They would walk right past the thriving local market, where small-scale business people of all kinds, from basket makers to vegetable growers to butchers to candy sellers, were displaying their entrepreneurial abilities in jobs they had created for themselves. Nathan spent his time talking to the people in the market, asking about their lives and businesses, learning what was in the way of those businesses expanding and incomes rising. He concluded that what was needed was not outside investors, but inside ones. Small loans available at reasonable interest rates, and classes in literacy and accounting, would produce much more long-term good for the community than bringing in a factory or assembly plant from outside.

Locate Responsibility in the System

That's a guideline both for analysis and design. In analysis, it means looking for the ways the system creates its own behavior. Do pay attention to the triggering events, the outside influences that bring forth one kind of behavior from the system rather than another. Sometimes those outside events can be controlled (as in reducing the pathogens in drinking water to keep down incidences of infectious disease). But sometimes they can't. And sometimes blaming or trying to control the outside influence blinds one to the easier task of increasing responsibility within the system.

"Intrinsic responsibility" means that the system is designed to send feedback about the consequences of decision making directly and quickly and compellingly to the decision makers. Because the pilot of a plane rides in the front of the plane, that pilot is intrinsically responsible. He or she will experience directly the consequences of his or her decisions.

Dartmouth College reduced intrinsic responsibility when it took thermostats out of individual offices and classrooms and put temperature-control decisions under the guidance of a central computer. That was done as an energy-saving measure. My observation from a low level in the hierarchy was that the main consequence was greater oscillations in room temperature. When my office got overheated, instead of turning down the thermostat, I had to call an office across campus, which got around to making corrections over a period of hours or days, and which often overcorrected, setting up the need for another phone call. One way of making that system more, rather than less, responsible might have been to let professors keep control of their own thermostats and charge them directly for the amount of energy they use, thereby privatizing a commons!

Designing a system for intrinsic responsibility could mean, for example, requiring all towns or companies that emit wastewater into a stream to place their intake pipes *downstream* from their outflow pipe. It could mean that neither insurance companies nor public funds should pay for medical costs resulting from smoking or from accidents in which a motorcycle rider didn't wear a helmet or a car rider didn't fasten the seat belt. It could mean Congress would no longer be allowed to legislate rules from which it exempts itself. (There are many rules from which Congress has exempted itself, including affirmative action hiring requirements and the necessity of preparing environmental impact statements.) A great deal of responsibility was lost when rulers who declared war were no longer expected to lead the troops into battle. Warfare became even more irresponsible when it became possible to push a button and cause tremendous damage at such a distance that the person pushing the button never even sees the damage.

Garrett Hardin has suggested that people who want to prevent other people from having an abortion are not practicing intrinsic responsibility, unless they are personally willing to bring up the resulting child![7]

These few examples are enough to get you thinking about how little our current culture has come to look for responsibility within the system that generates an action, and how poorly we design systems to experience the consequences of their actions.

Stay Humble—Stay a Learner

Systems thinking has taught me to trust my intuition more and my figuring-out rationality less, to lean on both as much as I can, but still to be prepared for surprises. Working with systems, on the computer, in nature, among people, in organizations, constantly reminds me of how incomplete my mental models are, how complex the world is, and how much I don't know.

The thing to do, when you don't know, is not to bluff and not to freeze, but to learn. The way you learn is by experiment—or, as Buckminster Fuller put it, by trial and error, error, error. In a world of complex systems, it is not appropriate to charge forward with rigid, undeviating directives. "Stay the course" is only a good idea if you're sure you're on course. Pretending you're in control even when you aren't is a recipe not only for mistakes, but for not learning from mistakes. What's appropriate when you're learning is small steps, constant monitoring, and a willingness to change course as you find out more about where it's leading.

That's hard. It means making mistakes and, worse, admitting them. It means what psychologist Don Michael calls "error-embracing." It takes a lot of courage to embrace your errors.

> Neither we ourselves, nor our associates, nor the publics that need to be involved . . . can learn what is going on and might go on if we act as if we really had the facts, were really certain about all the issues, knew exactly what the outcomes should/ could be, and were really certain that we were attaining the most preferred outcomes. Moreover, when addressing complex social issues, acting as if we knew what we were doing simply decreases our credibility. . . . Distrust of institutions and authority figures is increasing. The very act of acknowledging uncertainty could help greatly to reverse this worsening trend.[8]

> Error-embracing is the condition for learning. It means seeking and using—and sharing—information about what went wrong with what you expected or hoped would go right. Both error embracing and living with high levels of uncertainty emphasize our personal as well as societal vulnerability. Typically we hide our vulnerabilities from ourselves as well as from others. But . . . to be the kind of person who truly accepts his responsibility . . . requires knowledge of and access to self far beyond that possessed by most people in this society.[9]

Celebrate Complexity

Let's face it, the universe is messy. It is nonlinear, turbulent, and dynamic. It spends its time in transient behavior on its way to somewhere else, not in mathematically neat equilibria. It self-organizes and evolves. It creates diversity *and* uniformity. That's what makes the world interesting, that's what makes it beautiful, and that's what makes it work.

There's something within the human mind that is attracted to straight lines and not curves, to whole numbers and not fractions, to uniformity and not diversity, and to certainties and not mystery. But there is something else within us that has the opposite set of tendencies, since we ourselves evolved out of and are shaped by and structured as complex feedback systems. Only a part of us, a part that has emerged recently, designs buildings as boxes with uncompromising straight lines and flat surfaces. Another part of us recognizes instinctively that nature designs in fractals, with intriguing detail on every scale from the microscopic to the macroscopic. That part of us makes Gothic cathedrals and Persian carpets, symphonies and novels, Mardi Gras costumes and artificial intelligence programs, all with embellishments almost as complex as the ones we find in the world around us.

We can, and some of us do, celebrate and encourage self-organization, disorder, variety, and diversity. Some of us even make a moral code of doing so, as Aldo Leopold did with his land ethic: "A thing is right when it tends to preserve the integrity, stability, and beauty of the biotic community. It is wrong when it tends otherwise."[10]

Expand Time Horizons

One of the worst ideas humanity ever had was the interest rate, which led to the further ideas of payback periods and discount rates, all of which provide a rational, quantitative excuse for ignoring the long term.

The official time horizon of industrial society doesn't extend beyond what will happen after the next election or beyond the payback period of current investments. The time horizon of most families still extends farther than that—through the lifetimes of children or grandchildren. Many Native American cultures actively spoke of and considered in their decisions the effects on the seventh generation to come. The longer the operant time horizon, the better the chances for survival. As Kenneth Boulding wrote:

> There is a great deal of historical evidence to suggest that a society which loses its identity with posterity and which loses its positive image of the future loses also its capacity to deal with present problems, and soon falls apart. . . . There has always been something rather refreshing in the view that we should live like the birds, and perhaps posterity is for the birds in more senses than one; so perhaps we should all . . . go out and pollute something cheerfully. As an old taker of thought for the morrow, however, I cannot quite accept this solution. . . .[11]

In a strict systems sense, there is no long-term, short-term distinction. Phenomena at different time-scales are nested within each other. Actions taken now have some immediate effects and some that radiate out for decades to come. We experience now the consequences of actions set in motion yesterday and decades ago and centuries ago. The couplings between very fast processes and very slow ones are sometimes strong, sometimes weak. When the slow ones dominate, nothing seems to be happening; when the fast ones take over, things happen with breathtaking speed. Systems are always coupling and uncoupling the large and the small, the fast and the slow.

When you're walking along a tricky, curving, unknown, surprising, obstacle-strewn path, you'd be a fool to keep your head down and look just at the next step in front of you. You'd be equally a fool just to peer far ahead and never notice what's immediately under your feet. You need to be watching both the short and the long term—the whole system.

Defy the Disciplines

In spite of what you majored in, or what the textbooks say, or what you think you're an expert at, follow a system wherever it leads. It will be sure to lead across traditional disciplinary lines. To understand that system, you will have to be able to learn from—while not being limited by—economists and chemists and psychologists and theologians. You will have to penetrate their jargons, integrate what they tell you, recognize what they can honestly see through their particular lenses, and discard the distortions that come from the narrowness and incompleteness of their lenses. They won't make it easy for you.

Seeing systems whole requires more than being "interdisciplinary," if that word means, as it usually does, putting together people from different disciplines and letting them talk past each other. Interdisciplinary communication works only if there is a real problem to be solved, and if the representatives from the various disciplines are more committed to solving the problem than to being academically correct. They will have to go into learning mode. They will have to admit ignorance and be willing to be taught, by each other and by the system.

It can be done. It's very exciting when it happens.

Expand the Boundary of Caring

Living successfully in a world of complex systems means expanding not only time horizons and thought horizons; above all, it means expanding the horizons of caring. There are moral reasons for doing that, of course. And if moral arguments are not sufficient, then systems thinking provides the practical reasons to back up the moral ones. The real system is interconnected. No part of the human race is separate either from other human beings or from the global ecosystem. It will not be possible in this integrated world for your heart to succeed if your lungs fail, or for your company to succeed if your workers fail, or for the rich in Los Angeles to succeed if the poor in Los Angeles fail, or for Europe to succeed if Africa fails, or for the global economy to succeed if the global environment fails.

As with everything else about systems, most people already know about the interconnections that make moral and practical rules turn out to be the same rules. They just have to bring themselves to believe that which they know.

Don't Erode the Goal of Goodness

The most damaging example of the systems archetype called "drift to low performance" is the process by which modern industrial culture has eroded the goal of morality. The workings of the trap have been classic, and awful to behold.

Examples of bad human behavior are held up, magnified by the media, affirmed by the culture, as typical. This is just what you would expect. After all, we're only human. The far more numerous examples of human goodness are barely noticed. They are "not news." They are exceptions. Must have been a saint. Can't expect everyone to behave like that.

And so expectations are lowered. The gap between desired behavior and actual behavior narrows. Fewer actions are taken to affirm and instill ideals. The public discourse is full of cynicism. Public leaders are visibly, unrepentantly amoral or immoral and are not held to account. Idealism is ridiculed. Statements of moral belief are suspect. It is much easier to talk about hate in public than to talk about love. The literary critic and naturalist Joseph Wood Krutch put it this way:

Thus though man has never before been so complacent about what he *has*, or so confident of his ability to *do* whatever he sets his mind upon, it is at the same time true that he never before accepted so low an estimate of what *he is*. That same scientific method which enabled him to create his wealth and to unleash the power he wields has, he believes, enabled biology and psychology to explain him away—or at least to explain away whatever used to seem unique or even in any way mysterious. . . . Truly he is, for all his wealth and power, poor in spirit.[12]

We know what to do about drift to low performance. Don't weigh the bad news more heavily than the good. And keep standards absolute.

Systems thinking can only tell us to do that. It can't do it. We're back to the gap between understanding and implementation. Systems thinking by itself cannot bridge that gap, but it can lead us to the edge of what analysis can do and then point beyond—to what can and must be done by the human spirit.

Appendix

System Definitions: A Glossary

Archetypes: Common system structures that produce characteristic patterns of behavior.

Balancing feedback loop: A stabilizing, goal-seeking, regulating feedback loop, also know as a "negative feedback loop" because it opposes, or reverses, whatever direction of change is imposed on the system.

Bounded rationality: The logic that leads to decisions or actions that make sense within one part of a system but are not reasonable within a broader context or when seen as a part of the wider system.

Dynamic equilibrium: The condition in which the state of a stock (its level or its size) is steady and unchanging, despite inflows and outflows. This is possible only when all inflows equal all outflows.

Dynamics: The behavior over time of a system or any of its components.

Feedback loop: The mechanism (rule or information flow or signal) that allows a change in a stock to affect a flow into or out of that same stock. A closed chain of causal connections from a stock, through a set of decisions and actions dependent on the level of the stock, and back again through a flow to change the stock.

Flow: Material or information that enters or leaves a stock over a period of time.

Hierarchy: Systems organized in such a way as to create a larger system. Subsystems within systems.

Limiting factor: A necessary system input that is the one limiting the activity of the system at a particular moment.

Linear relationship: A relationship between two elements in a system that has constant proportion between cause and effect and so can be drawn with a straight line on a graph. The effect is additive.

Nonlinear relationship: A relationship between two elements in a system where the cause does not produce a proportional (straight-line) effect.

Reinforcing feedback loop: An amplifying or enhancing feedback loop, also known as a "positive feedback loop" because it reinforces the direction of change. These are vicious cycles and virtuous circles.

Resilience: The ability of a system to recover from perturbation; the ability to restore or repair or bounce back after a change due to an outside force.

Self-organization: The ability of a system to structure itself, to create new structure, to learn, or diversify.

Shifting dominance: The change over time of the relative strengths of competing feedback loops.

Stock: An accumulation of material or information that has built up in a system over time.

Suboptimization: The behavior resulting from a subsystem's goals dominating at the expense of the total system's goals.

System: A set of elements or parts that is coherently organized and interconnected in a pattern or structure that produces a characteristic set of behaviors, often classified as its "function" or "purpose."

Summary of Systems Principles

Systems

- A system is more than the sum of its parts.
- Many of the interconnections in systems operate through the flow of information.
- The least obvious part of the system, its function or purpose, is often the most crucial determinant of the system's behavior.
- System structure is the source of system behavior. System behavior reveals itself as a series of events over time.

Stocks, Flows, and Dynamic Equilibrium

- A stock is the memory of the history of changing flows within the system.
- If the sum of inflows exceeds the sum of outflows, the stock level will rise.
- If the sum of outflows exceeds the sum of inflows, the stock level will fall.
- If the sum of outflows equals the sum of inflows, the stock level will not change — it will be held in dynamic equilibrium.

- A stock can be increased by decreasing its outflow rate as well as by increasing its inflow rate.
- Stocks act as delays or buffers or shock absorbers in systems.
- Stocks allow inflows and outflows to be de-coupled and independent.

Feedback Loops

- A feedback loop is a closed chain of causal connections from a stock, through a set of decisions or rules or physical laws or actions that are dependent on the level of the stock, and back again through a flow to change the stock.
- Balancing feedback loops are equilibrating or goal-seeking structures in systems and are both sources of stability and sources of resistance to change.
- Reinforcing feedback loops are self-enhancing, leading to exponential growth or to runaway collapses over time.
- The information delivered by a feedback loop—even nonphysical feedback—can affect only future behavior; it can't deliver a signal fast enough to correct behavior that drove the current feedback.
- A stock-maintaining balancing feedback loop must have its goal set appropriately to compensate for draining or inflowing processes that affect that stock. Otherwise, the feedback process will fall short of or exceed the target for the stock.
- Systems with similar feedback structures produce similar dynamic behaviors.

Shifting Dominance, Delays, and Oscillations

- Complex behaviors of systems often arise as the relative strengths of feedback loops shift, causing first one loop and then another to dominate behavior.
- A delay in a balancing feedback loop makes a system likely to oscillate.
- Changing the length of a delay may make a large change in the behavior of a system.

Scenarios and Testing Models

- System dynamics models explore possible futures and ask "what if" questions.
- Model utility depends not on whether its driving scenarios are realistic (since no one can know that for sure), but on whether it responds with a realistic pattern of behavior.

Constraints on Systems

- In physical, exponentially growing systems, there must be at least one reinforcing loop driving the growth *and* at least one balancing loop constraining the growth, because no system can grow forever in a finite environment.
- Nonrenewable resources are stock-limited.
- Renewable resources are flow-limited.

Resilience, Self-Organization, and Hierarchy

- There are always limits to resilience.
- Systems need to be managed not only for productivity or stability, they also need to be managed for resilience.
- Systems often have the property of self-organization—the ability to structure themselves, to create new structure, to learn, diversify, and complexify.
- Hierarchical systems evolve from the bottom up. The purpose of the upper layers of the hierarchy is to serve the purposes of the lower layers.

Source of System Surprises

- Many relationships in systems are nonlinear.
- There are no separate systems. The world is a continuum. Where to draw a boundary around a system depends on the purpose of the discussion.
- At any given time, the input that is most important to a system is the one that is most limiting.
- Any physical entity with multiple inputs and outputs is surrounded by layers of limits.
- There always will be limits to growth.

- A quantity growing exponentially toward a limit reaches that limit in a surprisingly short time.
- When there are long delays in feedback loops, some sort of foresight is essential.
- The bounded rationality of each actor in a system may not lead to decisions that further the welfare of the system as a whole.

Mindsets and Models

- Everything we think we know about the world is a model.
- Our models do have a strong congruence with the world.
- Our models fall far short of representing the real world fully.

Springing the System Traps

Policy Resistance

Trap: When various actors try to pull a system state toward various goals, the result can be policy resistance. Any new policy, especially if it's effective, just pulls the system state farther from the goals of other actors and produces additional resistance, with a result that no one likes, but that everyone expends considerable effort in maintaining.

The Way Out: Let go. Bring in all the actors and use the energy formerly expended on resistance to seek out mutually satisfactory ways for all goals to be realized—or redefinitions of larger and more important goals that everyone can pull toward together.

The Tragedy of the Commons

Trap: When there is a commonly shared resource, every user benefits directly from its use, but shares the costs of its abuse with everyone else. Therefore, there is very weak feedback from the condition of the resource to the decisions of the resource users. The consequence is overuse of the resource, eroding it until it becomes unavailable to anyone.

The Way Out: Educate and exhort the users, so they understand the consequences of abusing the resource. And also restore or strengthen the missing feedback link, either by privatizing the resource so each user feels the direct consequences of its abuse or (since many resources cannot be privatized) by regulating the access of all users to the resource.

Drift to Low Performance

Trap: Allowing performance standards to be influenced by past performance, especially if there is a negative bias in perceiving past performance, sets up a reinforcing feedback loop of eroding goals that sets a system drifting toward low performance.

The Way Out: Keep performance standards absolute. Even better, let standards be enhanced by the best actual performances instead of being discouraged by the worst. Set up a drift toward high performance!

Escalation

Trap: When the state of one stock is determined by trying to surpass the state of another stock—and vice versa—then there is a reinforcing feedback loop carrying the system into an arms race, a wealth race, a smear campaign, escalating loudness, escalating violence. The escalation is exponential and can lead to extremes surprisingly quickly. If nothing is done, the spiral will be stopped by someone's collapse—because exponential growth cannot go on forever.

The Way Out: The best way out of this trap is to avoid getting in it. If caught in an escalating system, one can refuse to compete (unilaterally disarm), thereby interrupting the reinforcing loop. Or one can negotiate a new system with balancing loops to control the escalation.

Success to the Successful

Trap: If the winners of a competition are systematically rewarded with the means to win again, a reinforcing feedback loop is created by which, if it is allowed to proceed uninhibited, the winners eventually take all, while the losers are eliminated.

The Way Out: Diversification, which allows those who are losing the competition to get out of that game and start another one; strict limitation on the fraction of the pie any one winner may win (antitrust laws); policies that level the playing field, removing some of the advantage of the strongest players or increasing the advantage of the weakest; policies that devise rewards for success that do not bias the next round of competition.

Shifting the Burden to the Intervenor

Trap: Shifting the burden, dependence, and addiction arise when a solution to a systemic problem reduces (or disguises) the symptoms, but does nothing to solve the underlying problem. Whether it is a substance that dulls one's perception or a policy that hides the underlying trouble, the drug of choice interferes with the actions that could solve the real problem.

If the intervention designed to correct the problem causes the self-maintaining capacity of the original system to atrophy or erode, then a destructive reinforcing feedback loop is set in motion. The system deteriorates; more and more of the solution is then required. The system will become more and more dependent on the intervention and less and less able to maintain its own desired state.

The Way Out: Again, the best way out of this trap is to avoid getting in. Beware of symptom-relieving or signal-denying policies or practices that don't really address the problem. Take the focus off short-term relief and put it on long-term restructuring.

If you are the intervenor, work in such a way as to restore or enhance the system's own ability to solve its problems, then remove yourself.

If you are the one with an unsupportable dependency, build your system's own capabilities back up before removing the intervention. Do it right away. The longer you wait, the harder the withdrawal process will be.

Rule Beating

Trap: Rules to govern a system can lead to rule-beating—perverse behavior that gives the appearance of obeying the rules or achieving the goals, but that actually distorts the system.

The Way Out: Design, or redesign, rules to release creativity not in the direction of beating the rules, but in the direction of achieving the purpose of the rules.

Seeking the Wrong Goal

Trap: System behavior is particularly sensitive to the goals of feedback loops. If the goals—the indicators of satisfaction of the rules—are defined inaccurately or incompletely, the system may obediently work to produce a result that is not really intended or wanted.

The Way Out: Specify indicators and goals that reflect the real welfare of the system. Be especially careful not to confuse effort with result or you will end up with a system that is producing effort, not result.

Places to Intervene in a System
(in increasing order of effectiveness)

12. **Numbers:** Constants and parameters such as subsidies, taxes, and standards
11. **Buffers:** The sizes of stabilizing stocks relative to their flows
10. **Stock-and-Flow Structures:** Physical systems and their nodes of intersection
9. **Delays:** The lengths of time relative to the rates of system changes
8. **Balancing Feedback Loops:** The strength of the feedbacks relative to the impacts they are trying to correct
7. **Reinforcing Feedback Loops:** The strength of the gain of driving loops
6. **Information Flows:** The structure of who does and does not have access to information
5. **Rules:** Incentives, punishments, constraints
4. **Self-Organization:** The power to add, change, or evolve system structure
3. **Goals:** The purpose of the system
2. **Paradigms:** The mind-set out of which the system—its goals, structure, rules, delays, parameters—arises
1. **Transcending Paradigms**

Guidelines for Living in a World of Systems

1. Get the beat of the system.
2. Expose your mental models to the light of day.
3. Honor, respect, and distribute information.
4. Use language with care and enrich it with systems concepts.
5. Pay attention to what is important, not just what is quantifiable.
6. Make feedback policies for feedback systems.
7. Go for the good of the whole.

8. Listen to the wisdom of the system.

9. Locate responsibility within the system.

10. Stay humble—stay a learner.

11. Celebrate complexity.

12. Expand time horizons.

13. Defy the disciplines.

14. Expand the boundary of caring.

15. Don't erode the goal of goodness.

Model Equations

There is much to be learned about systems without using a computer. However, once you have started to explore the behavior of even very simple systems, you may well find that you wish to learn more about building your own formal mathematical models of systems. The models in this book were originally developed using *STELLA* modeling software, by isee systems Inc. (formerly High Performance Systems). The equations in this section are written to be easily translated into various modeling software, such as *Vensim* by Ventana Systems Inc. as well as *STELLA* and *iThink* by isee systems Inc.

The following model equations are those used for the nine dynamic models discussed in chapters 1 and 2. "Converters" can be constants or calculations based on other elements of the system model. Time is abbreviated (*t*) and the change in time from one calculation to the next, the time interval, is noted as (*dt*).

Chapter One

Bathtub—for Figures 5, 6 and 7
Stock: *water in tub(t)* = *water in tub(t − dt)* + (*inflow − outflow*) × *dt*
Initial stock value: *water in tub* = 50 gal
t = minutes
dt = 1 minute
Run time = 10 minutes
Inflow: *inflow* = 0 gal/min . . . for time 0 to 5; 5 gal/min . . . for time 6 to 10
Outflow: *outflow* = 5 gal/min

Coffee Cup Cooling or Warming—for Figures 10 and 11

Cooling

Stock: *coffee temperature(t)* = *coffee temperature(t − dt)* − *(cooling × dt)*

Initial stock value: *coffee temperature* = 100°C, 80°C, and 60°C . . . for three comparative model runs.

t = minutes

dt = 1 minute

Run time = 8 minutes

Outflow: *cooling* = *discrepancy* × 10%

Converters: *discrepancy* = *coffee temperature − room temperature*

room temperature = 18°C

Warming

Stock: *coffee temperature(t)* = *coffee temperature(t − dt)* + *(heating × dt)*

Initial stock value: *coffee temperature* = 0°C, 5°C, and 10°C . . . for three comparative model runs.

t = minutes

dt = 1 minute

Run time = 8 minutes

Inflow: *heating* = *discrepancy* × 10%

Converters: *discrepancy* = *room temperature − coffee temperature*

room temperature = 18°C

Bank Account—for Figures 12 and 13

Stock: *money in bank account(t)* = *money in bank account(t − dt)* + *(interest added × dt)*

Initial stock value: *money in bank account* = $100

t = years

dt = 1 year

Run time = 12 years

Inflow: *interest added ($/year)* = *money in bank account × interest rate*

Converter: *interest rate* = 2%, 4%, 6%, 8%, & 10% annual interest . . . for five comparative model runs.

Chapter Two

Thermostat—For Figures 14-20

Stock: *room temperature(t)* = *room temperature(t – dt)* + *(heat from furnace – heat to outside)* × *dt*

Initial stock value: *room temperature* = 10°C for cold-room warming; 18°C for warm-room cooling

t = hours

dt = 1 hour

Run time = 8 hours and 24 hours

Inflow: *heat from furnace* = minimum of *discrepancy between desired and actual room temperature* or 5

Outflow: *heat to outside* = *discrepancy between inside and outside temperature* × 10% . . . for "normal" house; *discrepancy between inside and outside temperature* × 30% . . . for very leaky house

Converters: *thermostat setting* = 18°C

discrepancy between desired and actual room temperature = maximum of (*thermostat setting – room temperature*) or 0

discrepancy between inside and outside temperature =
room temperature – 10°C . . . for constant outside temperature (Figures 16–18); *room temperature – 24-hour outside temp* . . . for full day-and-night cycle (Figures 19 and 20)

24-hour outside temp ranges from 10°C (50°F) during the day to – 5°C (23°F) at night, as shown in graph

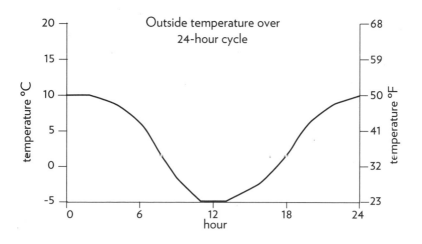

Population—for Figures 21–26
Stock: *population(t) = population(t − dt) + (births − deaths) × dt*
Initial stock value: *population* = 6.6 billion people
t = years
dt = 1 year
Run time = 100 years
Inflow: *births = population × fertility*
Outflow: *deaths = population × mortality*
Converters:

Figure 22:
mortality = .009 . . . or 9 deaths per 1000 population
fertility = .021 . . . or 21 births per 1000 population

Figure 23:
mortality = .030
fertility = .021

Figure 24:
mortality = .009
fertility starts at .021 and falls over time to .009 as shown in graph

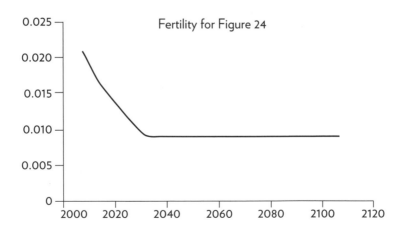

Figure 26:
mortality = .009
fertility starts at .021, drops to .009, but then rises .030 as shown in graph

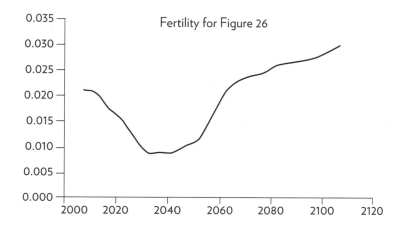

Fertility for Figure 26

Capital—for Figures 27 and 28
Stock: *capital stock(t) = capital stock(t − dt) + (investment − depreciation) × dt*
Initial stock value: *capital stock* = 100
t = years
dt = 1 year
Run time = 50 years
Inflow: *investment = annual output × investment fraction*
Outflow: *depreciation = capital stock / capital lifetime*
Converters: *annual output = capital stock × output per unit capital*
capital lifetime = 10 years, 15 years, and 20 years . . . for three comparative
   model runs.
investment fraction = 20%
output per unit capital = 1/3

Business Inventory—for Figures 29–36
Stock: *inventory of cars on the lot(t) =*
   inventory of cars on the lot(t − dt) + (deliveries − sales) × dt
Initial stock values: *inventory of cars on the lot* = 200 cars
t = days
dt = 1 day
Run time = 100 days

Inflows: *deliveries* = 20 . . . for time 0 to 5; *orders to factory (t – delivery delay)* . . . for time 6 to 100

Outflows: *sales* = minimum of *inventory of cars on the lot* or *customer demand*

Converters: *customer demand* = 20 cars per day . . . for time 0 to 25; 22 cars per day . . . for time 26 to 100

perceived sales = *sales* averaged over *perception delay* (i.e., *sales* smoothed over *perception delay*)

desired inventory = *perceived sales* × 10

discrepancy = *desired inventory* – *inventory of cars on the lot*

orders to factory = maximum of (*perceived sales* + *discrepancy*) or 0 . . . for Figure 32; maximum of (*perceived sales* + *discrepancy/response delay*) or 0 . . . for Figures 34–36

Delays, Figure 30:
perception delay = 0
response delay = 0
delivery delay = 0

Delays, Figure 32:
perception delay = 5 days
response delay = 3 days
delivery delay = 5 days

Delays, Figure 34:
perception delay = 2 days
response delay = 3 days
delivery delay = 5 days

Delays, Figure 35:
perception delay = 5 days
response delay = 2 days
delivery delay = 5 days

Delays, Figure 36:
perception delay = 5 days
response delay = 6 days
delivery delay = 5 days

A Renewable Stock Constrained by a Nonrenewable Resource—for Figures 37-41

Stock: *resource(t)* = *resource(t − dt)* − (*extraction* × *dt*)

Initial stock values: *resource* = 1000 . . . for Figures 38, 40, and 41; 1000, 2000, and 4000 . . . for three comparative model runs in Figure 39

Outflow: *extraction* = *capital* × *yield per unit capital*

t = years

dt = 1 year

Run time = 100 years

Stock: *capital(t)* = *capital(t − dt)* + (*investment − depreciation*) × *dt*

Initial stock values: *capital* = 5

Inflow: *investment* = minimum of *profit* or *growth goal*

Outflow: *depreciation* = *capital / capital lifetime*

Converters: *capital lifetime* = 20 years

profit = (*price* × *extraction*) − (*capital* × 10%)

growth goal = *capital* × 10% . . . for Figures 30–40; *capital* × 6%, 8%, 10%, and 12% . . . for four comparative model runs in Figure 40

price = 3 . . . for Figures 38, 39, and 40; for Figure 41, price starts at 1.2 when yield per unit capital is high and rises to 10 as yield per unit capital falls, as shown in graph

yield per unit capital starts at 1 when resource stock is high and falls to 0 as the resource stock declines, as shown in graph

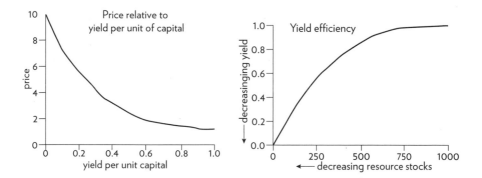

A Renewable Stock Constrained by a Renewable Resource—for Figures 42-45

Stock: $resource(t) = resource(t - dt) + (regeneration - harvest) \times dt$
Initial stock value: $resource = 1000$
Inflow: $regeneration = resource \times regeneration\ rate$
Outflow: $harvest = capital \times yield\ per\ unit\ capital$
t = years
dt = 1 year
Run time = 100 years

Stock: $capital(t) = capital(t - dt) + (investment - depreciation) \times dt$
Initial stock value: $capital = 5$
Inflow: $investment$ = minimum of $profit$ or $growth\ goal$
Outflow: $depreciation = capital\ /\ capital\ lifetime$

Converters: $capital\ lifetime = 20$
$growth\ goal = capital \times 10\%$
$profit = (price \times harvest) - capital$
$price$ starts at 1.2 when yield per unit capital is high and rises to 10 as
    yield per unit capital falls. This is the same nonlinear relationship for
    price and yield as in the previous model.

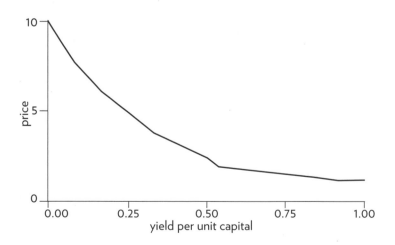

regeneration rate is 0 when the resource is either fully stocked or completely depleted. In the middle of the resource range, regeneration rate peaks near 0.5.

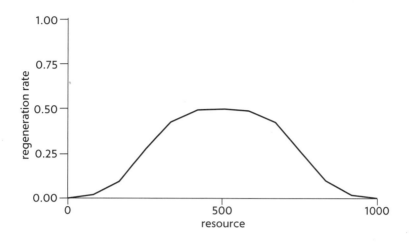

yield per unit capital starts at 1 when the resource is fully stocked, but falls (non-linearly) as the resource stock declines. Yield per unit capital increases overall from least efficient in Figure 43, to slightly more efficient in Figure 44, to most efficient in Figure 45.

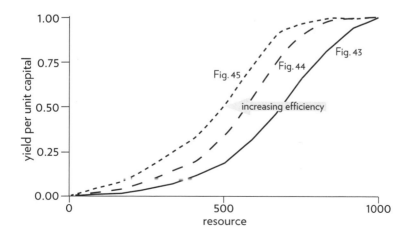

Notes

Introduction
1. Russell Ackoff, "The Future of Operational Research Is Past," *Journal of the Operational Research Society* 30, no. 2 (February 1979): 93–104.
2. Idries Shah, *Tales of the Dervishes* (New York: E. P. Dutton, 1970), 25.

Chapter One
1. Poul Anderson, quoted in Arthur Koestler, *The Ghost in the Machine* (New York: Macmillan, 1968), 59.
2. Ramon Margalef, "Perspectives in Ecological Theory," *Co-Evolution Quarterly* (Summer 1975), 49.
3. Jay W. Forrester, *Industrial Dynamics* (Cambridge, MA: The MIT Press, 1961), 15.
4. Honoré Balzac, quoted in George P. Richardson, *Feedback Thought in Social Science and Systems Theory* (Philadelphia: University of Pennsylvania Press, 1991), 54.
5. Jan Tinbergen, quoted in ibid, 44.

Chapter Two
1. Albert Einstein, "On the Method of Theoretical Physics," *The Herbert Spencer Lecture,* delivered at Oxford (10 June 1933); also published in *Philosophy of Science* 1, no. 2 (April 1934): 163–69.
2. The concept of a "systems zoo" was invented by Prof. Hartmut Bossel of the University of Kassel in Germany. His three recent "System Zoo" books contain system descriptions and simulation-model documentations of more than 100 "animals," some of which are included in modified form here. Hartmut Bossel, *System Zoo Simulation Models – Vol. 1: Elementary Systems, Physics, Engineering; Vol. 2: Climate, Ecosystems, Resources; Vol. 3: Economy, Society, Development.* (Norderstedt, Germany: Books on Demand, 2007).
3. For a more complete model, see the chapter "Population Sector" in Dennis L. Meadows et al., *Dynamics of Growth in a Finite World,* (Cambridge MA: Wright-Allen Press, 1974).
4. For an example, see Chapter 2 in Donella Meadows, Jørgen Randers, and Dennis Meadows, *Limits to Growth: The 30-Year Update* (White River Junction, VT: Chelsea Green Publishing Co., 2004).
5. Jay W. Forrester, 1989. "The System Dynamics National Model: Macrobehavior from Microstructure," in P. M. Milling and E. O. K. Zahn, eds., *Computer-Based Management of Complex Systems: International System Dynamics Conference* (Berlin: Springer-Verlag, 1989).

Chapter Three

1. Aldo Leopold, *Round River* (New York: Oxford University Press, 1993).
2. C. S. Holling, ed., *Adaptive Environmental Assessment and Management*, (Chichester UK: John Wiley & Sons, 1978), 34.
3. Ludwig von Bertalanffy, *Problems of Life: An Evaluation of Modern Biological Thought* (New York: John Wiley & Sons Inc., 1952), 105.
4. Jonathan Swift, "Poetry, a Rhapsody, 1733." In *The Poetical Works of Jonathan Swift* (Boston: Little Brown & Co.,1959).
5. Paraphrased from Herbert Simon, *The Sciences of the Artificial* (Cambridge MA: MIT Press, 1969), 90–91 and 98–99.

Chapter Four

1. Wendell Berry, *Standing by Words* (Washington, DC: Shoemaker & Hoard, 2005), 65.
2. Kenneth Boulding, "General Systems as a Point of View," in Mihajlo D. Mesarovic, ed., *Views on General Systems Theory*, proceedings of the Second Systems Symposium, Case Institute of Technology, Cleveland, April 1963 (New York: John Wiley & Sons, 1964).
3. James Gleick, *Chaos: Making a New Science* (New York: Viking, 1987), 23–24.
4. This story is compiled from the following sources: C. S. Holling, "The Curious Behavior of Complex Systems: Lessons from Ecology," in H. A. Linstone, *Future Research* (Reading, MA: Addison-Wesley, 1977); B. A. Montgomery et al., *The Spruce Budworm Handbook*, Michigan Cooperative Forest Pest Management Program, Handbook 82-7, November, 1982; *The Research News*, University of Michigan, April-June, 1984; Kari Lie, "The Spruce Budworm Controversy in New Brunswick and Nova Scotia," *Alternatives* 10, no. 10 (Spring 1980), 5; R. F. Morris, "The Dynamics of Epidemic Spruce Budworm Populations," *Entomological Society of Canada*, no. 31, (1963).
5. Garrett Hardin, "The Cybernetics of Competition: A Biologist's View of Society," *Perspectives in Biology and Medicine* 7, no. 1 (1963): 58-84.
6. Jay W. Forrester, *Urban Dynamics* (Cambridge, MA: The MIT Press, 1969), 117.
7. Václav Havel, from a speech to the Institute of France, quoted in the *International Herald Tribune*, November 13, 1992, p. 7.
8. Dennis L. Meadows, *Dynamics of Commodity Production Cycles*, (Cambridge MA: Wright-Allen Press, Inc., 1970).
9. Adam Smith, *An Inquiry into the Nature and Causes of the Wealth of Nations*, Edwin Cannan, ed., (Chicago: University of Chicago Press, 1976), 477-8.
10. Herman Daly, ed., *Toward a Steady-State Economy* (San Francisco: W. H. Freeman and Co., 1973), 17; Herbert Simon, "Theories of Bounded Rationality," in R. Radner and C. B. McGuire, eds., *Decision and Organization* (Amsterdam: North-Holland Pub. Co., 1972). ·
11. The term "satisficing" (a merging of "satisfy" and "suffice") was first used by Herbert Simon to describe the behavior of making decisions that meet needs adequately, rather than trying to maximize outcomes in the face of imperfect information. H. Simon, *Models of Man*, (New York: Wiley, 1957).
12. Philip G. Zimbardo, "On the Ethics of Intervention in Human Psychological Research: With Special Reference to the Stanford Prison Experiment," *Cognition* 2, no. 2 (1973): 243–56)
13. This story was told to me during a conference in Kollekolle, Denmark, in 1973.

Chapter Five

1. Paraphrased in an interview by Barry James, "Voltaire's Legacy: The Cult of the Systems Man," *International Herald Tribune*, December 16, 1992, p. 24.

2. John H. Cushman, Jr., "From Clinton, a Flyer on Corporate Jets?" *International Herald Tribune*, December 15, 1992, p. 11.

3. World Bank, *World Development Report 1984* (New York: Oxford University Press, 1984), 157; Petre Muresan and Ioan M. Copil, "Romania," in B. Berelson, ed., *Population Policy in Developed Countries* (New York: McGraw-Hill Book Company, 1974), 355-84.

4. Alva Myrdal, *Nation and Family* (Cambridge, MA: MIT Press, 1968). Original edition published New York: Harper & Brothers, 1941.

5. "Germans Lose Ground on Asylum Pact," *International Herald Tribune*, December 15, 1992, p. 5.

6. Garrett Hardin, "The Tragedy of the Commons," *Science* 162, no. 3859 (13 December 1968): 1243–48.

7. Erik Ipsen, "Britain on the Skids: A Malaise at the Top," *International Herald Tribune*, December 15, 1992, p. 1.

8. Clyde Haberman, "Israeli Soldier Kidnapped by Islamic Extremists," *International Herald Tribune*, December 14, 1992, p. 1.

9. Sylvia Nasar, "Clinton Tax Plan Meets Math," *International Herald Tribune*, December 14, 1992, p. 15.

10. See Jonathan Kozol, *Savage Inequalities: Children in America's Schools* (New York: Crown Publishers, 1991).

11. Quoted in Thomas L. Friedman, "Bill Clinton Live: Not Just a Talk Show," *International Herald Tribune*, December 16, 1992, p. 6.

12. Keith B. Richburg, "Addiction, Somali-Style, Worries Marines," *International Herald Tribune*, December 15, 1992, p. 2.

13. *Calvin and Hobbes* comic strip, *International Herald Tribune*, December 18, 1992, p. 22.

14. Wouter Tims, "Food, Agriculture, and Systems Analysis," *Options*, International Institute of Applied Systems Analysis Laxenburg, Austria no. 2 (1984), 16.

15. "Tokyo Cuts Outlook on Growth to 1.6%," *International Herald Tribune*, December 19-20, 1992, p. 11.

16. Robert F. Kennedy address, University of Kansas, Lawrence, Kansas, March 18, 1968. Available from the JFK Library On-Line, http://www.jfklibrary.org/Historical+Resources/Archives/Reference+Desk/Speeches/RFK/RFKSpeech68Mar18UKansas.htm. Accessed 6/11/08.

17. Wendell Berry, *Home Economics* (San Francisco: North Point Press, 1987), 133.

Chapter Six

1. Lawrence Malkin, "IBM Slashes Spending for Research in New Cutback," *International Herald Tribune*, December 16, 1992, p. 1.

2. J. W. Forrester, *World Dynamics* (Cambridge MA: Wright-Allen Press, 1971).

3. Forrester, *Urban Dynamics* (Cambridge, MA: The MIT Press, 1969), 65.

4. Thanks to David Holmstrom of Santiago, Chile.

5. For an example, see Dennis Meadows's model of commodity price fluctuations: Dennis L. Meadows, *Dynamics of Commodity Production Cycles* (Cambridge, MA: Wright-Allen Press, Inc., 1970).

6. John Kenneth Galbraith, *The New Industrial State* (Boston: Houghton Mifflin, 1967).

7. Ralph Waldo Emerson, "War," lecture delivered in Boston, March, 1838. Reprinted in *Emerson's Complete Works,* vol. XI, (Boston: Houghton, Mifflin & Co., 1887), 177.

8. Thomas Kuhn, *The Structure of Scientific Revolutions* (Chicago: University of Chicago Press, 1962).

Chapter Seven

1. G.K. Chesterton, *Orthodoxy* (New York: Dodd, Mead and Co., 1927).

2. For a beautiful example of how systems thinking and other human qualities can be combined in the context of corporate management, see Peter Senge's book *The Fifth Discipline: The Art and Practice of the Learning Organization* (New York: Doubleday, 1990).

3. Philip Abelson, "Major Changes in the Chemical Industry," *Science* 255, no. 5051 (20 March 1992), 1489.

4. Fred Kofman, "Double-Loop Accounting: A Language for the Learning Organization," *The Systems Thinker* 3, no. 1 (February 1992).

5. Wendell Berry, *Standing by Words* (San Francisco: North Point Press, 1983), 24, 52.

6. This story was told to me by Ed Roberts of Pugh-Roberts Associates.

7. Garrett Hardin, *Exploring New Ethics for Survival: the Voyage of the Spaceship Beagle* (New York, Penguin Books, 1976), 107.

8. Donald N. Michael, "Competences and Compassion in an Age of Uncertainty," *World Future Society Bulletin* (January/February 1983).

9. Donald N. Michael quoted in H. A. Linstone and W. H. C. Simmonds. eds., *Futures Research* (Reading, MA: Addison-Wesley, 1977), 98–99.

10. Aldo Leopold, *A Sand County Almanac and Sketches Here and There* (New York: Oxford University Press, 1968), 224–25.

11. Kenneth Boulding, "The Economics of the Coming Spaceship Earth," in H. Jarrett, ed., *Environmental Quality in a Growing Economy: Essays from the Sixth Resources for the Future Forum* (Baltimore, MD: Johns Hopkins University Press, 1966), 11-12.

12. Joseph Wood Krutch, *Human Nature and the Human Condition* (New York: Random House, 1959).

Bibliography of
Systems Resources

In addition to the works cited in the Notes, the items listed here are jumping off points—places to start your search for more ways to see and learn about systems. The fields of systems thinking and system dynamics are now extensive, reaching into many disciplines. For more resources, see also www.ThinkingInSystems.org

Systems Thinking and Modeling

Books

Bossel, Hartmut. *Systems and Models: Complexity, Dynamics, Evolution, Sustainability.* (Norderstedt, Germany: Books on Demand, 2007). A comprehensive textbook presenting the fundamental concepts and approaches for understanding and modeling the complex systems shaping the dynamics of our world, with a large bibliography on systems.

Bossel, Hartmut. *System Zoo Simulation Models.* Vol. 1: *Elementary Systems, Physics, Engineering;* Vol. 2: *Climate, Ecosystems, Resources;* Vol. 3: *Economy, Society, Development.* (Norderstedt, Germany: Books on Demand, 2007). A collection of more than 100 simulation models of dynamic systems from all fields of science, with full documentation of models, results, exercises, and free simulation model download.

Forrester, Jay. *Principles of Systems.* (Cambridge, MA: Pegasus Communications, 1990). First published in 1968, this is the original introductory text on system dynamics.

Laszlo, Ervin. *A Systems View of the World.* (Cresskill, NJ: Hampton Press, 1996).

Richardson, George P. *Feedback Thought in Social Science and Systems Theory.* (Philadelphia: University of Pennsylvania Press, 1991). The long, varied, and fascinating history of feedback concepts in social theory.

Sweeney, Linda B. and Dennis Meadows. *The Systems Thinking Playbook.* (2001). A collection of 30 short gaming exercises that illustrate lessons about systems thinking and mental models.

Organizations, Websites, Periodicals, and Software

Creative Learning Exchange—an organization devoted to developing "systems citizens" in K–12 education. Publisher of *The CLE Newsletter* and books for teachers and students. www.clexchange.org

isee systems, inc.—Developer of *STELLA* and *iThink* software for modeling dynamic systems. www.iseesystems.com

Pegasus Communications—Publisher of two newsletters, *The Systems Thinker* and *Leverage Points,* as well as many books and other resources on systems thinking. www.pegasuscom.com

System Dynamics Society—an international forum for researchers, educators, consultants, and practitioners dedicated to the development and use of systems thinking and system dynamics around the world. *The Systems Dynamics Review* is the official journal of the System Dynamics Society. www.systemdynamics.org

Ventana Systems, Inc.—Developer of *Vensim* software for modeling dynamic systems. vensim.com

Systems Thinking and Business

Senge, Peter. *The Fifth Discipline: The Art and Practice of the Learning Organization.* (New York: Doubleday, 1990). Systems thinking in a business environment, and also the broader philosophical tools that arise from and complement systems thinking, such as mental-model flexibility and visioning.

Sherwood, Dennis. *Seeing the Forest for the Trees: A Manager's Guide to Applying Systems Thinking.* (London: Nicholas Brealey Publishing, 2002).

Sterman, John D. *Business Dynamics: Systems Thinking and Modeling for a Complex World.* (Boston: Irwin McGraw Hill, 2000).

Systems Thinking and Environment

Ford, Andrew. *Modeling the Environment.* (Washington, DC: Island Press, 1999.)

Systems Thinking, Society, and Social Change

Macy, Joanna. *Mutual Causality in Buddhism and General Systems Theory.* (Albany, NY: Stat University of New York Press, 1991).

Meadows, Donella H. *The Global Citizen.* (Washington, DC: Island Press, 1991).

Editor's Acknowledgments

A great many people have helped bring this book to life. In her original manuscript, Donella (Dana) Meadows extended special thanks to the Balaton Group, the Environmental Systems Analysis Group at Kassel, the Environmental Studies Program at Dartmouth, Ian and Margo Baldwin and Chelsea Green Publishing, Hartmut and Rike Bossel, High Performance Systems (now known as isee systems), and many readers and commentators. She also noted the role of her extended "farm family," those people who, over the years, lived and worked on her organic farm in Plainfield, New Hampshire.

As the editor who readied Dana's manuscript for publication after her death, I would like to add more thanks: Ann and Hans Zulliger and the Foundation for the Third Millennium, along with the board and staff of the Sustainability Institute, have contributed support and enthusiasm to this project. Many advisors and reviewers have critiqued the text and models and helped me think through how to make this book useful to the world—Hartmut Bossel, Tom Fiddaman, Chris Soderquist, Phil Rice, Dennis Meadows, Beth Sawin, Helen Whybrow, Jim Schley, Peter Stein, Bert Cohen, Hunter Lovins, and the students at the Presidio School of Management. The entire team at Chelsea Green Publishing have crafted the complex manuscript into a clear book. I thank all of them for their work to help us be better stewards of our home planet.

And finally, I thank Dana Meadows for all that I have learned from her and through editing this book.

About the Author

Donella Meadows (1941–2001) was a scientist trained in chemistry and biophysics (Ph.D., Harvard University). In 1970, she joined the Massachusetts Institute of Technology team lead by Dennis Meadows that produced "World3," a global computer model that explores the dynamics of human population and economic growth on a finite planet. In 1972, she was lead author of *The Limits to Growth*, the book that described for the general public the insights from the World3 modeling project. *Limits* was translated into twenty-eight languages and sparked debate around the world about the earth's carrying capacity and human choices. Meadows went on to write nine more books on global modeling and sustainable development and for fifteen years she wrote a weekly column, "The Global Citizen," reflecting on the state of our society and the complex connections in the world.

In 1991, Meadows was recognized as a Pew Scholar in Conservation and the Environment, and in 1994 she received a MacArthur Fellowship. She founded the Sustainability Institute in 1996 to apply systems thinking and organizational learning to economic, environmental, and social challenges. From 1972 until her death in 2001, Meadows taught in the Environmental Studies Program of Dartmouth College.

Index

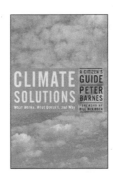

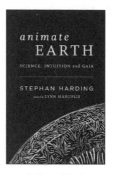